Advances in Intelligent and Soft Computing 72

Editor-in-Chief: J. Kacprzyk

Advances in Intelligent and Soft Computing

Editor-in-Chief

Prof. Janusz Kacprzyk
Systems Research Institute
Polish Academy of Sciences
ul. Newelska 6
01-447 Warsaw
Poland
E-mail: kacprzyk@ibspan.waw.pl

Further volumes of this series can be found on our homepage: springer.com

Vol. 56. H. Wang, Y. Shen, T. Huang, Z. Zeng (Eds.)
The Sixth International Symposium on Neural Networks (ISNN 2009), 2009
ISBN 978-3-642-01215-0

Vol. 57. M. Kurzynski, M. Wozniak (Eds.)
Computer Recognition Systems 3, 2009
ISBN 978-3-540-93904-7

Vol. 58. J. Mehnen, A. Tiwari, M. Köppen, A. Saad (Eds.)
Applications of Soft Computing, 2009
ISBN 978-3-540-89618-0

Vol. 59. K.A. Cyran, S. Kozielski, J.F. Peters, U. Stańczyk, A. Wakulicz-Deja (Eds.)
Man-Machine Interactions, 2009
ISBN 978-3-642-00562-6

Vol. 60. Z.S. Hippe, J.L. Kulikowski (Eds.)
Human-Computer Systems Interaction, 2009
ISBN 978-3-642-03201-1

Vol. 61. W. Yu, E.N. Sanchez (Eds.)
Advances in Computational Intelligence, 2009
ISBN 978-3-642-03155-7

Vol. 62. B. Cao, T.-F. Li, C.-Y. Zhang (Eds.)
Fuzzy Information and Engineering Volume 2, 2009
ISBN 978-3-642-03663-7

Vol. 63. Á. Herrero, P. Gastaldo, R. Zunino, E. Corchado (Eds.)
Computational Intelligence in Security for Information Systems, 2009
ISBN 978-3-642-04090-0

Vol. 64. E. Tkacz, A. Kapczynski (Eds.)
Internet – Technical Development and Applications, 2009
ISBN 978-3-642-05018-3

Vol. 65. E. Kącki, M. Rudnicki, J. Stempczyńska (Eds.)
Computers in Medical Activity, 2009
ISBN 978-3-642-04461-8

Vol. 66. G.Q. Huang, K.L. Mak, P.G. Maropoulos (Eds.)
Proceedings of the 6th CIRP-Sponsored International Conference on Digital Enterprise Technology, 2009
ISBN 978-3-642-10429-9

Vol. 67. V. Snášel, P.S. Szczepaniak, A. Abraham, J. Kacprzyk (Eds.)
Advances in Intelligent Web Mastering - 2, 2010
ISBN 978-3-642-10686-6

Vol. 68. V.-N. Huynh, Y. Nakamori, J. Lawry, M. Inuiguchi (Eds.)
Integrated Uncertainty Management and Applications, 2010
ISBN 978-3-642-11959-0

Vol. 69. E. Piętka and J. Kawa (Eds.)
Information Technologies in Biomedicine, 2010
ISBN 978-3-642-13104-2

Vol. 70. XXX

Vol. 71. XXX

Vol. 72. J.C. Augusto, J.M. Corchado, P. Novais, C. Analide (Eds.)
Ambient Intelligence, Future Trends - International Symposium on Ambient Intelligence (ISAmI 2010)
ISBN 978-3-642-13267-4

Juan Carlos Augusto, Juan M. Corchado,
Paulo Novais, and Cesar Analide (Eds.)

Ambient Intelligence and Future Trends-International Symposium on Ambient Intelligence (ISAmI 2010)

Springer

Editors

Juan C. Augusto
School of Computing
and Mathematics
University of Ulster at Jordanstown
Newtownabbey
United Kingdom
E-mail: jc.augusto@ulster.ac.uk

Paulo Novais
Universidade do Minho
Departamento de Informática
Campus de Gualtar
4710-057 Braga
Portugal
Email: pjon@di.uminho.pt

Juan M. Corchado
Departamento de Informática
y Automática
Facultad de Ciencias
Universidad de Salamanca
Plaza de la Merced S/N
37008, Salamanca
Spain
E-mail: corchado@usal.es

Cesar Analide
Universidade do Minho
Departamento de Informática
Campus de Gualtar
4710-057 Braga
Portugal
E-mail: analide@di.uminho.pt

ISBN 978-3-642-13267-4 e-ISBN 978-3-642-13268-1

DOI 10.1007/978-3-642-13268-1

Advances in Intelligent and Soft Computing ISSN 1867-5662

Library of Congress Control Number: 2010927159

Typeset & Cover Design: Scientific Publishing Services Pvt. Ltd., Chennai, India.

Printed on acid-free paper

5 4 3 2 1 0

springer.com

Preface

Ambient Intelligence (AmI) is a recent paradigm, emerging from the confluence of several areas like Artificial Intelligence and Ubiquitous Computing, where computers are used as proactive tools assisting people in their day-to-day to make their lives more comfortable. The interaction with computers is changing quickly and becoming more natural for us, this is an important aim for AmI emphasizing interaction with computational systems through friendly interfaces, like those based on natural language or simple gestures. This inclusion of technology in our day-to-day objects and environments should be as invisible as possible, which can be facilitated through the computational power and communication technologies embedded in many devices available nowadays.

ISAmI is the International Symposium on Ambient Intelligence, aiming to bring together researchers from various disciplines that constitute the scientific field of Ambient Intelligence to present and discuss the latest results, new ideas, projects and lessons obtained from recent experiences in building AmI systems.

This volume presents the papers that have been accepted in this first edition. These papers reports on innovative results and advances achieved recently in this area. Each paper has been reviewed by three different reviewers, from an international committee composed of 58 members from 18 different countries. From the 52 submissions received, 21 were selected for full presentation at the conference, 9 were accepted as short papers and 4 papers accept in the doctoral consortium.

We would like to thank all the contributing authors, as well as the members of the Program Committee and the Local Organizing Committee for their hard and highly valuable work. Their work has helped to contribute to the success of the ISAmI 2010 event. Thanks for your help, ISAmI2010 would not have existed without your contribution.

June 2010

The editors
Juan C. Augusto
Juan M. Corchado
Paulo Novais
Cesar Analide

Organization

Scientific Committee Chairs

Juan Carlos Augusto	University of Ulster (UK)
Juan M. Corchado	University of Salamanca (Spain)

Organizing Committee Chairs

Paulo Novais	University of Minho (Portugal)
Cesar Analide	University of Minho (Portugal)

Program Committe

Aaron Quigley	Human Interface Technology Laboratory (Australia)
Alexander Meschtscherjakov	University of Salzburg (Austria)
Alireza Sahami	Univeristy of Duisburg Essen (Germany)
András Lörincz	Eotvos Lorand University (Hungary)
Andreas Riener	Johannes Kepler University Linz (Austria)
Antonio F. Gómez Skarmeta	University of Murcia (Spain)
Artur Lugmayr	Tampere Univ. of Technology (Finland)
Carlos Bento	University of Coimbra (Portugal)
Carlos Juiz	University of the Balearic Islands (Spain)
Carlos Ramos	Polytechnic of Porto (Portugal)
Cecilio Angulo	Polytechnic University of Catalonia (Spain)
Celia Gutierrez	Complutense University of Madrid (Spain)
Clemens Holzmann	Johannes Kepler University Linz (Austria)
Cristina Buiza	Ingema (Spain)
Dante I. Tapia	University of Salamanca (Spain)
Davy Preuveneers	Katholieke Universiteit Leuven (Belgium)
Diego Gachet	European University of Madrid (Spain)
Eduardo Dias	University of Évora (Portugal)
Elisabeth Eichhorn	Potsdam University of Applied Sciences (Germany)

Emilio S. Corchado	University of Burgos (Spain)
Francisco C. Pereira	University of Coimbra (Portugal)
Francisco J. Ballesteros	Rey Juan Carlos University (Spain)
Francisco Moya	University of Castilla-La Mancha (Spain)
Guenter Haring	University of Vienna (Austria)
Goreti Marreiros	Polytechnic of Porto (Portugal)
Gregor Broll	DOCOMO Euro-Labs (Germany)
Guillaume Lopez	University of Tokyo (Japan)
Hani Hagras	University of Essex (UK)
Hannu Korhonen	Nokia Research (Finland)
Helena Rodrigues	Universidade do Minho (Portugal)
Javier Bajo	Pontifical University of Salamanca (Spain)
Javier Jaen	Polytechnic University of Valencia (Spain)
Jesús Favela	CICESE (Mexico)
Jo Vermeulen	Hasselt University (Belgium)
Jose Bravo	University of Castilla-La Mancha (Spain)
José Danado	YDreams (Portugal)
José M. Molina	University Carlos III of Madrid (Spain)
José Machado	University of Minho (Portugal)
José Neves	University of Minho (Portugal)
Joyca Lacroix	Philips Research (Netherlands)
Kostas Stathis	University of London (UK)
Kristof Van Laerhoven	TU Darmstadt (Germany)
Lourdes Borrajo	University of Vigo (Spain)
Mariano Alcañiz	Polytechnic University of Valencia (Spain)
Martijn Vastenburg	Delft University of Technology (Netherlands)
Norbert Streitz	Smart Future Initiative (Germany)
Oliver Amft	TU Eindhoven (Netherlands)
Owen Noel Newton Fernando	National University of Singapore (Singapore)
Rene Meier	Trinity College Dublin (Ireland)
Rubén Fuentes Fernández	Complutense University of Madrid (Spain)
Rui José	University of Minho (Portugal)
Sara Rodríguez	University of Salamanca (Spain)
Sergio Ochoa	University of Chile (Chile)
Teresa Romão	New University of Lisbon (Portugal)
Victor M. Gonzalez	Universidad Autónoma de Nuevo León (México)
Wolfgang Reitberger	University of Salzburg (Austria)
Woontack Woo	GIST (Korea)
Yang Cai	Carnegie Mellon University (USA)

Local Organization Committee

Ângelo Costa	University of Minho (Portugal)
Cesar Analide	University of Minho (Portugal)
Davide Carneiro	University of Minho (Portugal)

Luís Lima	Polytechnic of Porto (Portugal)
Luís Machado	University of Minho (Portugal)
Paulo Novais	University of Minho (Portugal)
Pedro Gomes	University of Minho (Portugal)
Ricardo Costa	Polytechnic of Porto (Portugal)
Sara Fernandes	University of Minho (Portugal)

Contents

Long Papers

Short Papers

Doctoral Consortium

A Study on Autonomic Decision Method for Smart Gas Environments in Korea

Jeong Seok Oh, Jang Sik Park, and Jeong Rock Kwon

Abstract. Ubiquitous technologies are used in the latest industry trend, the information analysis paradigm shifts to smart service environments. The smart service includes autonomic operations in order to define the status of industry facilities. Furthermore, information analysis based on IT used to frequently data mining for detecting the meaningful information and deriving new pattern. This paper suggests decision method by analyzing automatically the status information in city gas facilities in order to service smart gas safety management. We modify data algorithm for fitting the domain of gas safety, construct decision model by the proposed algorithm, and demonstrate our method. As the accuracy of our method is improved over 90%, our approach can apply to smart gas safety management based on ubiquitous environments.

Keywords: Smart Gas service, ubiquitous technologies, decision model.

1 Introduction

The information analysis paradigm might shift to smart service environment, as ubiquitous technologies have been introduced in the latest industry trend [7,8]. Several organizations have been pushing convergence among heterogeneous fields using ubiquitous technologies, since the government of Korea made announcement on ubiquitous strategies such as u-it 839, u-Korea and u-City [7, 8]. The more convergence among various fields might be invigorated, the more heterogeneous data are rapidly generated through facilities, buildings, person, and so on. Ubiquitous technologies are capable of developing smart environment and service by analyzing the collected status information, although it is necessary to use development of sensor nodes and construction of network infrastructures [2]. Furthermore, the intellectualization of smart industry environment might include not only automatic operation but also autonomic operation. Autonomic operations are able to manage self-activity process. Based on the accumulated data, several

Jeong Seok Oh · Jang Sik Park · Jeong Rock Kwon
Institute of Gas Safety R&D, Korea Gas Safety Corporation, Korea
e-mail: jsoh90@gmail.com, {pjsik,jrkwon}@kgs.or.kr

J.C. Augusto et al. (Eds.): ISAmI 2010, AISC 72, pp. 1–9.
springerlink.com

studies have been developed risk analysis methods which might create the meaningful information, associate inter-relationship among information, and derive new pattern and information [4, 5, 11]. Especially, data mining method can make the information decision model using the accumulated data, and is capable of classifying the new data into specific criterion.

Many technologies and researches with priority given to automation have been studied in industry environment like as city gas. For example, GASNET(Gasline Network Sensor System) could provide automatic management by means of real-time communication and data collection about the status data on gas pipeline[10]. However, it is inadequate to try technologies and researches on autonomic system, which is necessary to smart processing according to analysis and decision of the status information on industry facilities.

This paper is to propose autonomic decision method for self-classifying the accumulated status information on gas facilities using data mining method[1]. Firstly, we are to collect real environment data in gas facilities. The real environment data are divided by values of protective potential on text-boxes of pipelines and values of third-party pipelines. Secondly, we derive the necessity of improving decision algorithm for using smart gas environment and propose approach by using data reduction and instance based processing[4, 5, 6]. Finally, we modify data mining algorithm depending on data domain of gas facilities. This approach is able to create another class of instance using clustering, and classify through changing the similarity computation method using coefficient of variation[13, 3]. The object of this research can support cornerstone in order to aid autonomic decision model for smart safety management related gas facilities in Korea.

2 Collecting Status Data

This research collected real status data on city gas facilities for effective decision method. The one of status data are values of protective potential on text-box of pipelines. These values are consist of maximum, minimum and average electric potential and are received with mobile communication in a few city gas companies of Korea. The monitoring system of some city gas company decides the status of pipeline according to the fixed threshold values. The threshold is differently defined per the city-gas company, but it is used same threshold on all pipeline of a company. Table 1 shows these status data which are gained about 6,420 instances for four months.

Table 1 The status data example of protective potential on test boxes of gas pipelines

Date	Maximum	Minimum	Average	Status
2009/10/29 17:06	-1554	-1372	-1455	Normal
2009/06/27 20:55	-879	-741	-818	Notice
2009/07/16 14:44	-161	-150	-150	Warning

The other of status data are measurable values on third-party gas pipelines such as under bridges. These measurable values are consist of temperature, humidity, intensity of illumination and infrared lights data and received with Zigbee multi-hop sensor network. We developed monitoring program in order to gather the status data on sensor network and collected about 18,000 instances in underground, inner, outer environment. Example of these data is shown Table 2.

Table 2 The status data example in third-party pipelines

Temperature	Humidity	Intensity of illumination	Infrared lights	Measured places
8	60	55	125	Underground
7	42	770	167	Underground
24	34	803	150	Inner
25	33	792	141	Inner
14	24	1003	196	Outer
8	30	990	186	Outer

Furthermore, possible values of industry facilities might be complicated and similar range. Especially, this situation shows frequently when the distance is close between facilities. Therefore, this research generates more complicated the status data by means of considering subdivision of measured places as shown Table 3.

Table 3 The status data example of subdivision of measured places

Temperature	Humidity	Intensity of illumination	Infrared lights	Measured places
8	60	55	125	Underground_1
7	42	770	167	Underground+3
24	34	803	150	Inner_5
25	33	792	141	Inner_2
14	24	1003	196	Outer_3
8	30	990	186	Outer_7

3 The Necessity of Improving the Decision Approach

In order to predict and analyze new data in data mining, classification methods are generally used. This research adopts J48 algorithm. The J48 algorithm, which is modified version of C4.5 algorithm, processes not only numeric data but also nominal data, and creates decision tree as a result model. The decision model is validated whether it is the reasonable accuracy. In general, the model is used in application when over 90% of accuracy. We generate the status model in test boxes and third-party pipelines, and verify the model using 10-cross validation method. Fig. 1 shows tree model based on the collected protective potential data in test boxes of underground pipelines. Instance attributes are maximum, minimum and average electric potential except date, and class names are normal, notice, and warning. This model can be used for classifying and predicting the status of test-boxes in underground pipelines, as it is 99.98% of accuracy using 10-cross validation method.

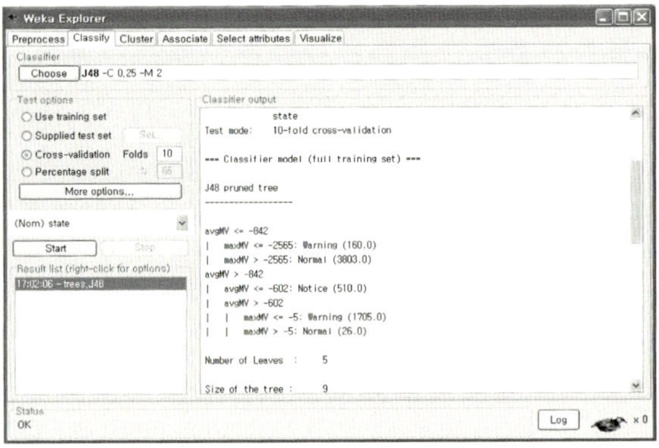

Fig. 1 The status decision model for protective potential

Furthermore, we create the status decision model based on the collected data in third-party pipelines/ Instance attributes are temperature, humidity, intensity of illumination, and infrared light. Class names are inner, underground, and outer. This model shows simple trees, too. The underground is less than 171V in infrared light and over 39% in humidity. The inner is less than 171V in infrared light and less than 39% in humidity. The outer is over 171V in infrared light. This model can be also used for self-classifying the status of third-party pipelines, as it is 99.97% of accuracy using 10-cross validation method.

However, generated status values in real environment might be more complicated. For example, if it is necessary to identify the location of nodes in our environment, class name is more classified in detail. Therefore, we subdivide class name that are described per sensor node such as "place_number of sensor node" as partially shown Fig. 2. This model shows rather complicated trees with only 206 leaf nodes, and then is difficult for self-deciding only classification method because it is 88.02% of accuracy using 10-corss validation method.

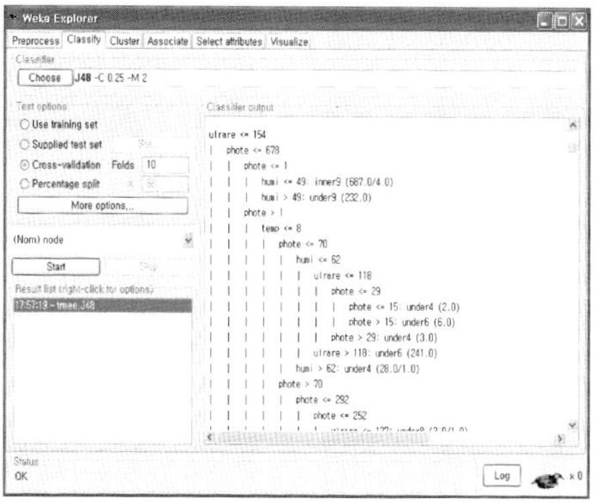

Fig. 2 The decision status model with class name in detail

4 The Autonomic Decision Method for Smart City Gas

The classification method might do not use as it stands, if data set include complicated and various range because of over-fitting problems. The over-fitting problem can happen when all data are too subdivided by class. In general, tree pruning or threshold of error rate is capable of solving this problem, but has not effect on our environment. Therefore, we propose the necessity of approach by using data reduction and instance based processing. The data reduction is data cube aggregation, dimension reduction, data compression, numerosity reduction and discretization and hierarchy generation. The numerosity reduction changes or estimates from whole data to small data set by means of parametric and non-parametric model such as histogram, clustering and sampling. This research is to focus methodology for autonomic decision on clustering method in numerosity reduction.

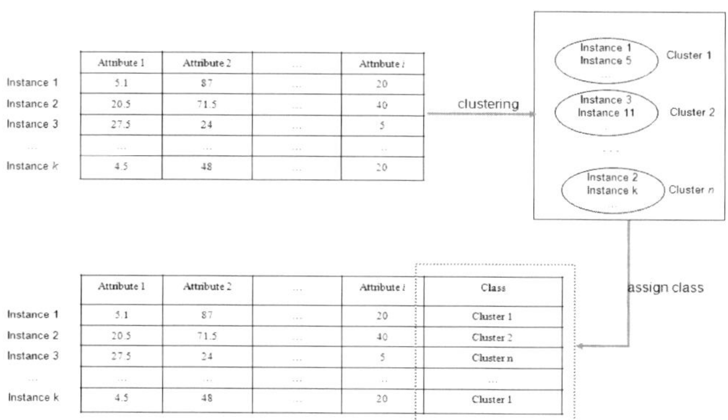

Fig. 3 Procedures with clustering

First step of our approach binds up similar data into a meaningful group in order to process efficiently whole data using clustering. Clustering might solve overfitting problem by generating automatically meaningful groups. Meaningful groups assign class name of instances, individually. Fig. 3 shows assigning procedures of collected instances using clustering method. Instance set consist of *i* attributes and *k* instances. Clustering generates *n* clusters with high similarity among instances. Cluster names are assigned as class name of instance. For example, the class name of instance 1 is cluster 1 because instance 1 exists in that cluster

In general, clustering algorithm might process the related instance set as a class unit. These processing results are inadequate to analyze results in detail since they regard several instances as a group. This group might be not able to show a typical characteristic of instance. In other words, it is necessary that the classification and prediction method can detect most similar instance in a cluster when new the status data input. Therefore, Second step of our approach creates another class of instance in order to find the most similar instance in a cluster. Fig. 4 schematizes another class as mapping information between instance and cluster in order to recommend most similar instance. This approach might be possible to provide the autonomic analysis service to the location of node of environment when second class is the location of node.

	Attribute 1	Attribute 2	...	Attribute i	First class	Second class
Instance 1	5.1	87	...	20	Cluster 1	Under_node_2
Instance 2	20.5	71.5	...	40	Cluster 2	Inner_node_8
Instance 3	27.5	24	...	5	Cluster n	Outer_node_3
...	...	...	...	...	...	...
Instance k	4.5	48	...	20	Cluster 1	Inner_node_2

Fig. 4 Mapping information example

Finding the most similar instance is capable of computing the similarity among instances in a cluster. The computation of similarity uses commonly Euclidian and Manhattan methods. However, these methods are not applied to our research with multi-dimensional data, which make the biased result due to various value ranges in multi-dimensional data. Final step of our approach changes the similarity computation method using coefficient of variation so that any attribute in multi-dimensional instance should not influence classification or prediction results. The similarity is computed by distance between instances. The similarity between new instance and an instance in cluster more decreases, both instances are more similar.

Fig. 5 shows processes of the similarity between new instance and instances in a first class using Euclidian, Manhattan and our method. All instances have 9 dimensions and each dimension have different value range. Euclidian and Manhattan methods compute instance 1 as the lowest distance and then recommend

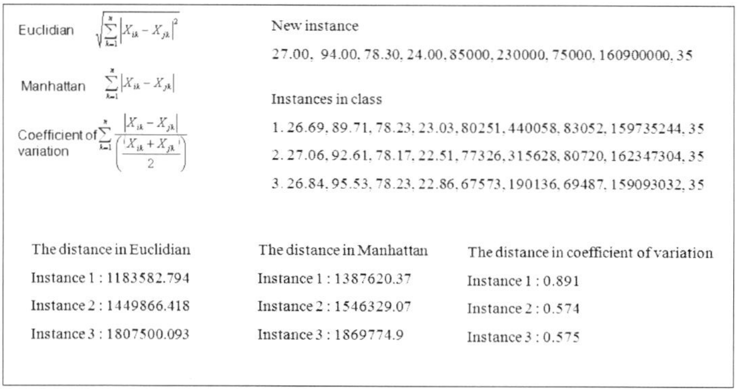

Fig. 5 Similarity computation procedures

instance 1 as the most similar instance to new instance. These methods process equally all attributes without regard for the different range of each attribute. However, our method computes and recommend instance 2 as the lowest distance to new instance by means of reducing the bias on different value range. We develop the proposed method by modifying public WEKA software.. This model can be also used for self-analyzing the status of third-party pipelines, as it is 99.91% of accuracy using 10-cross validation method. Our model has been played a basic role for classifying information in intelligent information layer, as shown Fig. 6.

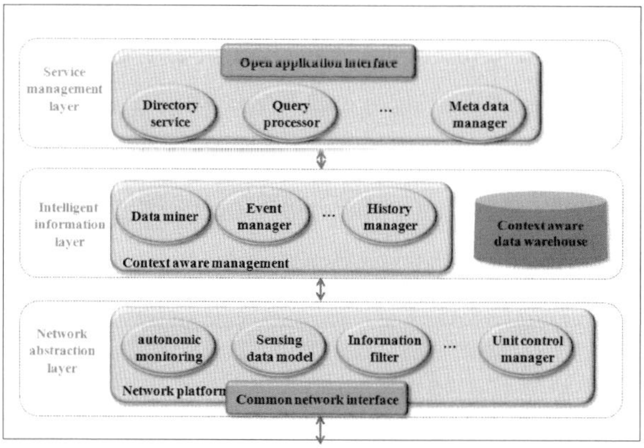

Fig. 6 Middleware for smart city gas environment

5 Conclusions and Future Works

The information analysis paradigm might shift to smart service environment, as ubiquitous technologies have been introduced in the latest industry trend. Furthermore, the intellectualization of smart industry environment might include not only automatic operation but also autonomic operation. Especially, data mining method can make information decision model using the accumulated data, and classify the new generated data into specific criterion. However, it is inadequate to try technologies and researches on autonomic system, which is necessary to smart processing according to analysis and decision of the status information on industry facilities.

This paper proposed autonomic decision method for self-classifying the accumulated status information on gas facilities using data mining method. Firstly, we collect real environment data in gas facilities. The real environment data are divided by values of protective potential on text-boxes of pipelines and values of third-party pipelines. The instance of protective potential consists of maximum, minimum and average electronic potential and is saved about 6.420 instances through mobile communication. The instance of third-party pipeline consists of temperature, humidity, intensity of illumination and infrared light, and is gathered about 18,000 instances through Zigbee multi-hop sensor network.

Secondly, we derived the necessity of improving decision algorithm for using smart gas environment. It is difficult to analyze only classification because of over-fitting problems. In general, tree pruning or threshold of error rate is capable of solving this problem, but has not effect on our environment. Therefore, we proposed the need of approach by using data reduction and instance based processing.

Finally, we modified data mining algorithm depending on data domain of gas facilities. Our method binds up similar data into a meaningful group using clustering, and then creates another class of instance unit in order to find the most similar in a cluster. Also, we change the similarity computation method using coefficient of variation for the purpose of considering multi-dimensional attributes, and test accuracy of the status decision mode that use our method with n-cross validation method. As future works, we are to develop smart city-gas management middleware that might include directory service, autonomic decision, context-aware service, and so on.

Acknowledgements. This work was supported by MIKE (Ministry of Knowledge Economy) under the program of ETI (Energy Technology Innovation). This paper is a result of "Research Group of Energy Safety for Next Generation". (Project NO 2007-M-CC23-P-03-1-000).

References

[1] Ganek, A., Corbi, T.: The Dawning of the Autonomic Computing Era. IBM systems Journal 42, 5–18 (2003)

[2] Culler, D.E., Hong, W.: Wireless Sensor Network. Communications of the ACM 47, 30–33 (2004)

[3] Moorel, D.: Statistics Concepts and Controversies, 5th edn. W.H.,Freeman and Company, New York (2001)

[4] Zangle, G., Hannerer, J.: Data Mining Applications in Petroleum Industry. IBM Oak Publishing (2003)

[5] Witten, H., Franke, E.: Data Mining Practical Machine Learning Tools and Techniques with Java Implements. Morgan Kaufmann Publisher, San Francisco (1999)
[6] Yu, H., Yang, J., Han, J.: Classifying Large Data Sets using SVM with Hierarchical Clusters. In: ACM SIGKDD Conference, pp. 306–315 (2003)
[7] Oh, J.S., Park, J.S., Kwon, J.R.: Design Middleware Platforms Smart Service on City Gas Environment in Korea. Communications in Computer and Information Science 62, 90–97 (2009)
[8] Oh, J.S., Park, J.S., Kwon, J.R.: Selecting the Wireless Communication Methods for Establishing Ubiquitous City Gas Environment in Korea. LNCS, vol. 5576, pp. 823–828. Springer, Heidelberg (2009)
[9] Goeble, M., Gruenwald, L.: A Survey of Data Mining and Knowledge Discovery Software Tools. ACM SIGKDD Explorations 1, 20–33 (1999)
[10] Schempf, H.: GasNet Project Final Report. AutoMatika, Inc. (2006)
[11] Elnaffar, S., Martin, P., Horma, R.: Automatically Classifying Database Workloads. In: CKIM Conference, pp. 622–624 (2002)
[12] Mitchell, T.M.: Machine Laerning. McGraw-Hill Companies, Inc., New York (1997)
[13] Weka Public Machine Learning Software (version 3.4), http://www.cs.waikato.ac/nz/weka (accessed May 2, 2005)

Multiagent Systems and Wearable Devices: Helping People Live Healthier

Juan A. Fraile, Carolina Zato, Oscar Gil, and Fernando de la Prieta

Abstract. This paper proposes a multi-agent system (MAS) that uses smart wearable devices and mobile technology for the care of patients in a geriatric home care facility. The use of RFID and NFC technologies allows remote monitoring and identification of patients, and makes it possible for them to receive treatment according to preventative medical protocol. The proposed multi-agent system manages the infrastructure of services within the environment both efficiently and securely by reasoning, task-planning and synchronizing the data obtained from the sensors. A system prototype was installed in a real environment and the results obtained are presented in this paper.

Keywords: Wearable Systems, Human Health, Dependent Environment, Smart Wearable Devices.

1 Introduction

The world population is aging and that both the number of people suffering from chronic illnesses and the elderly population in general are rapidly increasing. New technologies, wearable devices and communication networks are helping residents and their caregivers by providing continuous medical supervision, home automation control, rapid access to medical data, and emergency communication both in the home and in geriatric care facilities [1] [8]. During the last several years there has been an increase in the demand for non-invasive intelligent devices that can

Juan A. Fraile
Pontifical University of Salamanca, c/ Compañía 5, 37002 Salamanca, Spain
e-mail: jafraileni@upsa.es

Carolina Zato · Oscar Gil · Fernando de la Prieta
Departamento de Informática y Automática, University of Salamanca,
Plaza de la Merced s/n, 37008 Salamanca, Spain
e-mail: {carol_zato,oscar.gil,fer}@usal.es

J.C. Augusto et al. (Eds.): ISAmI 2010, AISC 72, pp. 11–18.

detect vital signs without interfering in the daily routine of those who use them [13]. Smart Wearable devices are sensors that are designed to help and provide services to dependent persons [9]. They offer new possibilities and create new challenges as a result of the information that they supply quickly, reliably and securely. As wireless technologies and ubiquitous computing [1] continue to unfold, network sensors [12], communication data networks [3], mobile devices and smart wearable devices [13] make it possible to create intelligent environments and assist with long-distance medical care [8] [12].

Agents and multi-agent systems have become basic components in the development of dynamic and distributed systems, and have been successfully applied in areas such as electronic commerce, medicine, homecare, robotics, etc. [4] [5] [6] [7]. The characteristics of the agents make them suitable for developing WSN (Wireless Sensor Network) based distributed systems. The advantage of the agents is their ability to easily adapt to the users and the characteristics of the surrounding environment [11]. Additionally, they are frequently used with mobile technology such as GPRS (General Packet Radio Service), NFC (Near Field Communication) or Bluetooth, making it possible for them to obtain context-aware information in novel ways.

This article describes the SWeDe (Smart Wearable Device) multi-agent system which is intended to automatically manage and improve the assistance to patients in geriatric facilities by using smart wearable technology. This system uses smart wearable devices to obtain context aware information, thus making it possible to identify, locate and gather data for each patient. The agents are in charge of managing and controlling the information obtained by the system. The core agent in SWeDe is the Head Agent, which is integrated into the system. The purpose of this agent is to improve the quality of life for the user by providing efficient and relevant solutions in execution time. The communication between the agents and the smart wearable devices is conducted with wireless technologies: Wi-Fi, ZigBee [3], NFC [14], while Radiofrequency (RFID) [12] is used for identification [15]. These technologies provide the structure that is required for supporting the communication needs for the system agents with devices and data handling equipment.

The remainder of the article is structured as follows: the next Section describes the proposed multi-agent system, focusing on the description of how the agents interact with the smart wearable devices. Section 3 presents a prototype that was developed for the care of patients in a geriatric facility. Finally, Section 4 presents the results and a conclusion obtained after installing the prototype in the care facility, and suggests future areas of investigation for improving the system.

2 SWeDe Multiagent System

Multi-agent systems are distributed systems based on the cooperation of multiple autonomous agents [4]. SWeDe is a multi-agent system that has evolved from the Thomas architecture [2] to facilitate the integration of agents and smart wearable devices via wireless networks and mobile technology. Thomas [2] is a multiagent system architecture that is useful for the new generation of virtual organizations in open environments. It consists of agents that are capable of autonomous decision making, incorporate learning mechanisms, and are able to respond to events by planning and replanning in execution time. Thomas is an open architecture that can easily incorporate any type of agent. The SWeDe multi-agent system is a specialized feature of the Thomas architecture for intelligent environments that can address the need to improve techniques for obtaining resident and patient data, as well as assign diagnoses in hospital centers and geriatric facilities, and monitor all types of patients. The SWeDe multi-agent system is a distributed agent platform that uses a ZigBee WSN to establish remote communication between patients and caregivers.

All smart wearable devices in SWeDe are based on RFID technology. These devices, which identify and contain information for the residents, are hidden within their clothes and uniforms so as to not inconvenience the residents in any way. RFID readers and mobile devices with NFC technology are used to read the RFID tags. NFC is a protocol based on a short range wireless interface. NFC fuses the usability of the RFID technology chip with the portability and market penetration of mobile telephony. It uses an electromagnetic field to communicate with a mobile telephone, PDA or laptop using an intelligent RFID tag, or with other devices to perform micro payments, exchange of information and access control. Additionally, the patients wear a series of sensors throughout the body that form a BAN (Body Area Network) with which it is possible to obtain information about the patient's state of health [10]. These sensor devices are very small and comfortable so that the patients do not reject them. At the same time, energy consumption should be optimal to avoid having to replace the battery so often. The SWeDe system not only gathers and stores user information, but it is also a proactive computing system since it anticipates user needs. The use of proactive computing [10] and smart wearable devices allows SWeDe to provide the user with a degree of independence, and enriches the quality of life for both patients and caregivers.

The SWeDe multi-agent system is based on a BDI (Belief, Desire, Intention) model [4] [5] in which the agents function as controllers and coordinators for various medical care tasks. The agents can initiate services on demand, or according to planned actions or user requests. The behavior of each agent can change with any modification made to a patient's diagnosis. The same agents can modify patient treatment and respond to the smart wearable devices installed in the environment. Additionally, the information on the state of the system is continuously stored in order allow for recovery in the event of system failure. The most important characteristics of the system are: (i) the Head Agent has reasoning capability; it can analyze and reason the context data gathered by the system and provide proactive solutions, (ii) the Head Agent can easily adapt to the context within which it acts and (iii) gather sensor data and messages from other agents in order to provide efficient solutions.

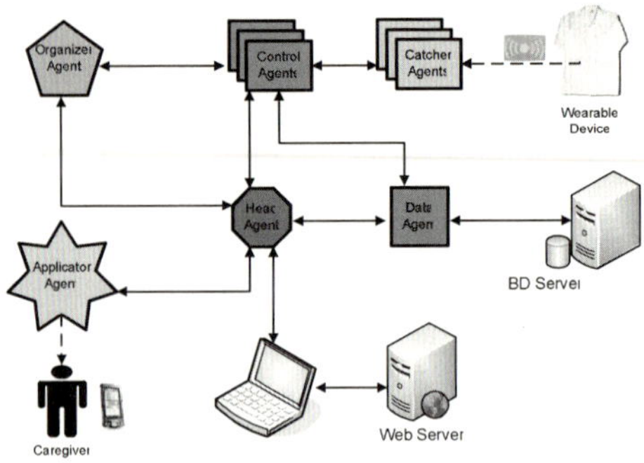

Fig. 1 SWeDe system agents

The different types of agents shown in Figure 1 are:

- Catcher agents: manage the portable RFID readers in order to obtain information from the RFID tags. There is one Catcher agent for each type of information (location, identification, security, access control, etc) that is captured in the system. There is also a Catcher agent installed in mobile devices with NFC technology to gather information from the RFID tags, which it sends to other agents via Bluetooth, GPRS or the ZigBee WSN.
- Control agents: supervise the activities of each of the Catcher agents with the help of control directives, and determine the validity of the information that is provided. They also classify the information received from the Catcher agents, organize it and send it to the Data agent and Organizer agent. Moreover, they administer the plans generated by the Organizer agent and ensure the plans are carried out by sending the information to the Catcher agents. Finally, they inform the Head agent about all of the incidents that arise during their processing.
- Data agent: structures and stores the information that it receives from the Control Agents in the information system. This effort facilitates the work of the Organizer agent. The stored information can be consulted via the Internet.
- Organizer agent: automatically generates supervision plans and simple patient diagnoses. These plans are sent to the Head agent, which is charge of distributing them between the Control agent and the Applicator agent.
- Head agent: controls the rest of the system agents. Manages the activation or deactivation of the Control agents. Also receives the incident reports from the other agents (database errors, non-executed plans, security notifications, connections and disconnections by the Control agents and Catcher agents…) The Head agent sends this information (incident, date, time, system state, device affected, detecting agent) to the Data agent to be registered in the information system. Additionally, the Head agent is responsible for informing the Applicator agent about these incidents, according to their priority level.
- Applicator agent: administers the communication with the caregivers and medical personnel. Receives incident reports from the Head agent, as well as information on the supervision tasks that have been carried out with the patients. All of this information can be sent via Bluetooth, Wi-Fi, ZigBee or GPRS. In the

event of any communication error, the Head agent is informed so that the information can be present at a later time.

The Organizer agent in SWeDe integrates reasoning (CBR) and planning (CBP) mechanisms, which allows the agents to use previous experiences to create better plans and achieve their objectives. Case-based Reasoning (CBR) [4] is a type of reasoning based on previous experiences. CBR systems solve new problems by adapting solutions that have been used to solve similar problems in the past, and learn from the new experiences. Case-based Planning (CBP) is specifically designed to generate plans (action sequences) [4] from past experiences in execution time. In CBP, the proposal for solving a problem is a plan. This solution is generated after taking into consideration the plans that have been applied to solve similar problems in the past. The problems and their respective plans are stored in a memory of plans. The reasoning mechanism generates the plans that it wants to use from past experiences and planning strategies, which is how the concept of Case-based Planning is understood [5].

3 Case Study: Geriatric Residences Monitoring

The present case study developed a multi-agent SWeDe system prototype in the San Juan de Sahagún (Salamanca) geriatric care facility, which improves the quality of life for the patients while simplifying and automating the work of the Personal Health Assistants (PHA) in a geriatric care facility. This system maintains continual identification and location of the patients and PHA within the dependent environment. Additionally, it automatically carries out planning and task distribution among the PHA. The information that the system gathers from the sensors, and the actions carried out by the patients on a daily basis are all stored and processed by the system in order to be able to provide a diagnosis or alert of possible incidents at a later time.

The system uses mobile devices (PDA, mobile telephones, NFC) and wireless communication technology (WiFi, GPRS, RFID, NFC and ZigBee) to interact with the users. These technologies and device work in a distributed way, providing the users with a greater flexibility and easier access to the system services. The standard ZigBee [3] permits more than 65,000 nodes to be connected in a network mesh topology. The ZigBee end devices remain in sleep mode in order to save energy and maximize their battery life. The RFID technology [12] used in the development of this prototype allows us to automatically capture data that can be used to electronically identify patients, caregivers and material resources. A RFID system is essentially comprised of 2 components: tags and readers. The RFID tag reader, which can be a mobile phone with NFC technology, gathers information as it moves throughout the tag's reading area, which it then transmits to the system. This information is comprised of the identification for the RFID tag that is being read, the identification of the reader that is being used, and the information contained in the RFID tag. The NFC mobiles transmit this information to the system via Bluetooth or GPRS, while the RFID readers use WiFi to transmit the information via the internal network installed within the context.

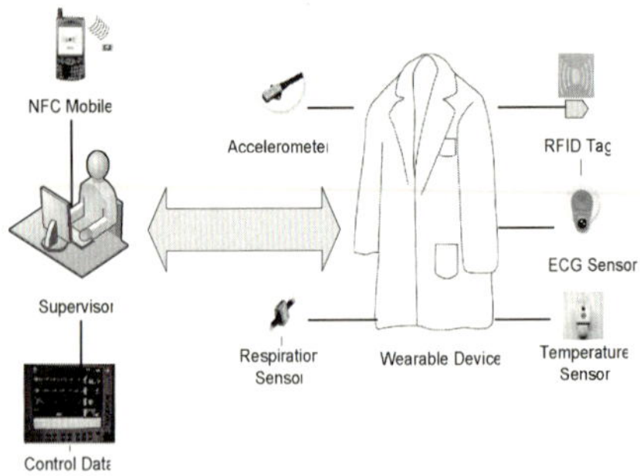

Fig. 2 Characteristics of the wearable device

Figure 2 illustrates the clothing used by the patients in the case study. It incorporates wearable devices, a RFID tag, an ECG sensory, a body temperature sensor, an accelerometer and a respiration sensor. Each of these devices transmits the information gathered from the patient to the control system via the ZigBee WSN. The RFID tag locates the patients and identifies them in the access control for the dining area, entertainment rooms, security areas and bedrooms. The ECG sensor is applied with a strip of Velcro to an area of the patient's clothing close enough to measure the heartbeat with an electric frequency. The signals received by the ECG can diagnose cardiovascular diseases or metabolic alterations. A triple axis accelerometer gathers signals on the patient's movements. The body temperature sensor monitors the patient's temperature. This sensor is an electronic device with a long life battery. It gathers new patient data every 60 seconds. The respiration sensor is another type of electronic device with a long life battery that also gathers information every 60 seconds. This particular frequency synchronizes perfectly with the definition of the respiratory frequency, such as the number of breaths per minute, and offers a satisfactory resolution of frequency and precision.

SWeDe is a specialized feature of the Thomas architecture that introduces compatibility with smart wearable devices through the use of ZigBee WSN and RFID and NFC technologies in order to establish remote communication between patients and caregivers. The primary characteristics of SWeDe are the modeling capabilities for intelligent environments based on virtual organizations, the incorporation of special agents capable of interacting with smart wearable devices, the incorporation of agents with learning and adaptation capabilities, and the integration of a large variety of wireless technologies.

4 Results and Conclusions

Smart wearable technology [13] is particularly promising as a support for health care environments. Sensors and communication devices facilitate sending and receiving information in a ubiquitous and non-intrusive manner [16]. The SWeDe system presented in this paper proposes an innovative technological solution consisting of an intelligent environment based on the use of smart wearable devices and multiagent systems [5]. The SWeDe system has been tested in a real environment, and the results obtained are promising. To validate the system, we designed a series of tests. The tests, which involved 50 patients, 10 nurses and 2 doctors, allowed us to evaluate the system. Specifically, the results of the SWeDe system were studied over a period of 5 weeks.

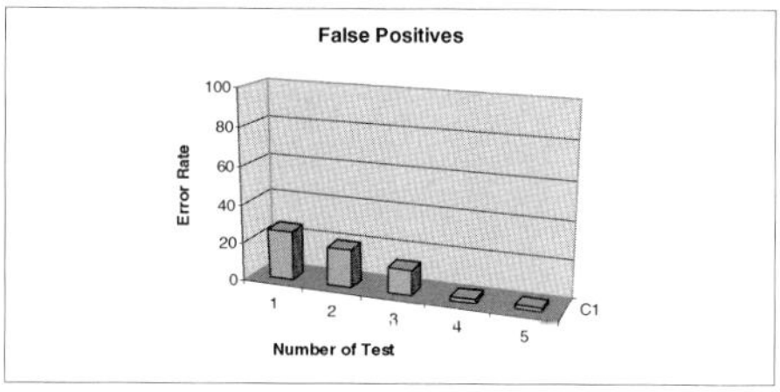

Fig. 3 False positives of the SWeDe system

Figure 3 shows the percentage of error that occurred in the escape control system during the tests. The initial tests showed an error rate above 26%. This percentage was primarily due to errors in the use of the system by the nurses. After the third test, the error rate was reduced to 13%. From that point the error rate remained stable at 2%. This error could be reduced with a higher level of training in the use of the system. Furthermore, as previously indicated, the Organizer agent uses a CBR strategy, and CBR systems typically require a period of time to learn from the environment, before providing accurate results.

Although the initial results are promising, the systems still requires much work. The SWeDe system has been successful tested in a real environment and has proved to be an efficient monitoring mechanism that uses context-aware techniques based on multi-agent systems and smart wearable devices. Our future work focuses on testing the system in different environments and looking for feedback to adjust and improve the proposed solution. That is our next challenge.

Acknowledgements. This work has been supported by the Spanish Ministry of Science and Innovation TIN 2009-13839-C03-03.

References

1. Abowd, G.D., Hayes, G.R., Iachello, G., Kientz, J.A., Patel, S.N., Stevens, M.M., Truong, K.N.: Prototypes and paratypes: Designing mobile and ubiquitous computing applications. IEEE Pervasive Computing, 67–73 (2005)
2. Bajo, J., Fraile, J.A., Corchado, J.M., Pérez-Lancho, B.: The THOMAS Architecture in Home Care Scenarios: A case study. Expert Systems with Applications 37(5), 3986–3999 (2010)
3. Baronti, P., Pillai, P., Chook, V.W.C., Chessa, S., Gotta, A., Hu, Y.F.: Wireless sensor networks: A survey on the state of the art and the 802.15.4 and ZigBee standards. Computer Communications 30(7), 1655–1695 (2007)

4. Corchado, J.M., Bajo, J., de Paz, Y., Tapia, D.: Intelligent Environment for Monitoring Alzheimer Patients. Agent Technology for Health Care. Decision Support Systems 34(2), 382–396 (2008)
5. Corchado, J.M., Bajo, J., Abraham, A.: GERAmI: Improving the delivery of health care. IEEE Intelligent Systems. Special Issue on Ambient Intelligence 32(2), 19–25 (2008)
6. Fraile, J.A., Bajo, J., Corchado, J.M.: Applying Context-Aware Computing in Dependent Environments. In: Mira, J., Ferrández, J.M., Álvarez, J.R., de la Paz, F., Toledo, F.J. (eds.) IWINAC 2009. LNCS, vol. 5602, pp. 85–94. Springer, Heidelberg (2009)
7. Fraile, J.A., Bajo, J., Corchado, J.M.: Multi-Agent Architecture for Dependent Environments. In: Providing Solutions for Home Care. Inteligencia Artificial. Special Issue 7th Ibero-American Workshop in Multi-Agent Systems, vol. 42, pp. 36–45 (2009)
8. Freitas, A.A., Wieser, D.C., Apweiler, R.: On the importance of comprehensible classification models for protein function prediction. IEEE/ACM Transactions on Computational Biology and Bioinformatics (2008)
9. Giroux, S., Pigot, H.: From Smart Homes to Smart Care: ICOST 2005 (Assistive Technology & Research). IOS Press, USA (2005)
10. Hsu, Y., Yang, C., Tsai, T., Cheng, C., Wu, C.: Development of a Decentralized Home Telehealth Monitoring System. Telemedicine and e-Health 13(1), 69–78 (2007)
11. Jayaputera, G.T., Zaslavsky, A.B., Loke, S.W.: Enabling run-time composition and support for heterogeneous pervasive multi-agent systems. Journal of Systems and Software 80(12), 2039–2062 (2007)
12. Lin, C.C., Lin, P.Y., Lu, P.K., Hsieh, G.Y., Lee, W.L., Lee, R.G.: A Healthcare Integration System for Disease Assessment and Safety Monitoring of Dementia Patients. IEEE Transactions on Information Technology in Biomedicine 12(5), 579–586 (2008)
13. Paradiso, J., Borriello, G., Bonato, P.: Implantable Electronics. IEEE Pervasive Computing 7(1), 12–13 (2008)
14. Strömmer, E., Kaartinen, J., Pärkkä, J., Ylisaukkooja, A., Korhonen, I.: Application of Near Field Communication for Health Monitoring in Daily life. In: Proceedings of 28th IEEE EMBS International Conference, New York City, pp. 3246–3249 (2006)
15. Want, R.: An Introduction to RFID Technology. IEEE Pervasive Computing 5, 25–33 (2006)
16. Yan, Y.S., Zhang, Y.T.: An Efficient Motion-Resistant Method for Wearable Pulse Oximeter. IEEE Transactions on Information Technology in Biomedicine 12(3), 399–405 (2008)

Accurate Temporal Relationships in Sequences of User Behaviours in Intelligent Environments

Asier Aztiria, Juan Carlos Augusto, Rosa Basagoiti, and Alberto Izaguirre

Abstract. Intelligent Environments are supposed to act proactively anticipating users' needs and preferences in order to provide effective support. Therefore, learning users' frequent behaviours is essential to provide such personalized services. In that sense, we have developed a system, which learns those frequent behaviours. Due to the complexity of the entire learning system, this paper will focus on discovering accurate temporal relationships to define the relationships between actions of the user.

Keywords: Ambient Intelligence, Learning patterns, Temporal relations.

1 Introduction

Ambient Intelligence (AmI) [1] [2] [3] can be understood as 'a digital environment that proactively, but sensibly, supports people in their daily lives' [4]. Some of the potential benefits that this technology can bring to people in their daily lives include making an environment more comfortable, safer and more energy efficient. In order to achieve these objectives, the environment should learn patterns of the user which means that the environment has to gain knowledge about the preferences, needs and habits of the user in order to be in a better position to assist the user adequately [5].

Let us consider the following scenario, which exemplifies a common behaviour of a user. *On weekdays Michael's alarm clock goes off ('Alarm on') few minutes after 08:00am. Approximately 10 minutes after getting up he usually steps into the*

Asier Aztiria · Rosa Basagoiti · Alberto Izaguirre
University of Mondragon
e-mail: aaztiria@eps.mondragon.edu, rbasagoiti@eps.mondragon.edu, aizagirre@eps.mondragon.edu

Juan Carlos Augusto
University of Ulster
e-mail: jc.augusto@ulster.ac.uk

J.C. Augusto et al. (Eds.): ISAmI 2010, AISC 72, pp. 19–27.

bathroom ('Bathroom on') and (4 seconds after) he turns on the lights of the bathroom ('BathLights on') if the bathroom is dark (bathroom light level <10). On Tuesdays, Thursdays and Fridays he usually takes a shower ('Shower on'); Michael prefers the temperature of the water to be around 24-26 degrees Celsius in the winter and around 21-23 degrees Celsius in the summer. Before he leaves the bathroom ('Bathroom off') he turns off the lights ('BathLights off').

Knowing users' frequent behaviours allows the environment to act intelligently and proactively. In Michael's case, it could mean that the environment automatically turns on and off the lights, sets the temperature of the water and so on. Unlike previous systems, *automation* of actions and/or devices in intelligent environments is based on learned patterns, making sure they adapt to users' common behaviours. Knowledge extracted from these patterns can also be used in order to *understand* his behaviour. For example, the analysis of frequent interaction with objects and devices in the house can facilitate the detection of unhealthy habits (e.g., the system detects that Michael does not brush his teeth in the mornings). Making the environment more efficient in terms of *saving energy* (e.g. by turning off the lights when he leaves) or *increasing safety* (e.g. turning off the water or issuing alarms when detecting that Michael left it on and he will not return soon) are other dimensions of daily life that can be supported by the Intelligent Environment thanks to the knowledge it has collected.

In order to achieve these objectives, we have developed a software which allows an Intelligent Environment to discover frequent behavioural patterns, and we have named it Sequential Patterns of User Behaviour System (SPUBS). Due to the complexity of SPUBS, this paper will focus on the step of discovering accurate temporal relationships between actions. The reminder of the paper is organized as follows. Section 2 describes briefly the SPUBS. Section 3 summarizes previous related work done in learning highlighting the task of discovering temporal relationships. In Section 4 we explain our approach for discovering accurate temporal relationships. Finally Section 5 shows the results of the validation experiments and Section 6 the conclusions.

2 Sequential Patterns of User Behaviour System

Sequential Patterns of User Behaviour (SPUBS) is a system that discovers user's common behaviours and habits from data recorded by sensors. The core of the system is the learning algorithm $\mathscr{A}_{SPUBS}$, which combined with a language $\mathscr{L}_{SPUBS}$, allows to discover and represent the patterns. This algorithm is divided in four steps which represent the four logical steps to discover frequent and comprehensible patterns. A brief description of each module follows:

Identifying Frequent Sets: The objective of this first step is to identify the sets of actions that frequently occur together (for further details see [6]). In Michael's scenario this step discovers that Michael usually carries out the following actions together:

'Alarm on', 'Bathroom on', 'Bathroom off', 'BathLights on', 'BathLights off', 'Shower on' and 'Shower off'

Identifying Topology: This step discovers in what order the user usually carries out the sets of actions discovered in the first step, giving it a sense of sequence (for further details see [7]). Thus, in Michael's scenario it discovers that the activation of the alarm clock comes first, then he goes into the bathroom, then he turns the lights on and so on. See Figure 1(a) for the output achieved after this step.

Identifying Time Relations: The objective of this step is to relate the actions in terms of time. In contrast with the previous step this step tries to relate the actions as accurately as possible (for further details see Section 4). See Figure 1(b) for the output achieved after this step.

Identifying Conditions: The last step of the algorithm is to discover specific conditions for situations where different options are, and general conditions in order to contextualize the whole sequence. In Michael's scenario, on the one hand it defines that he only has a shower on Tuesdays, Thursdays and Fridays. On the other hand this step also defines that the whole sequence occurs only on weekdays and it starts few minutes after 08:00am and it finishes before 09:00am. See Figure 1(c) for the output achieved after this step.

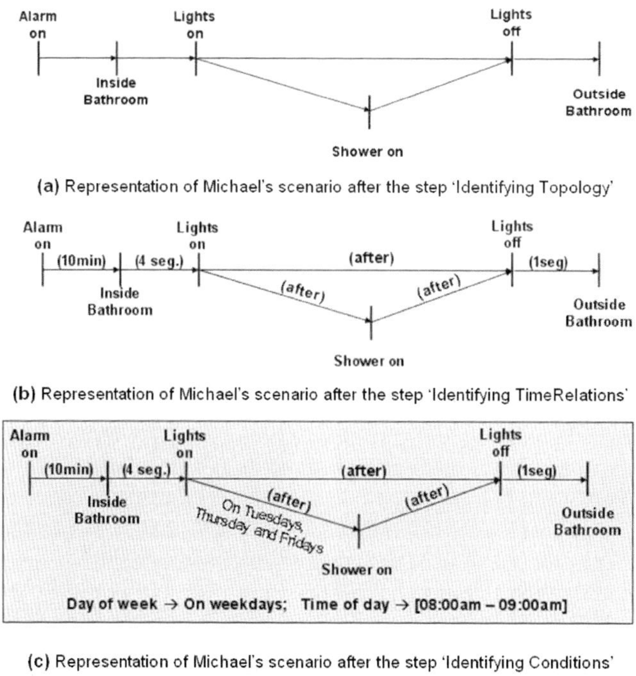

(a) Representation of Michael's scenario after the step 'Identifying Topology'

(b) Representation of Michael's scenario after the step 'Identifying TimeRelations'

(c) Representation of Michael's scenario after the step 'Identifying Conditions'

Fig. 1 Evolution of Michael's scenario through the $\mathscr{A}_{SPUBS}$

Although the whole algorithm is already developed, because of space restrictions, this paper will be focused on the first step of 'Identifying Time Relations'.

3 Related Work

Learning is an essential feature in any Intelligent Environment. However, it has not been devoted as much attention in the literature as it may require. Next, different approaches are analysed emphasizing the way they discover and represent the temporal relationships between actions.

So far, most of the applications [8] [9] [10] where a learning process was involved have related an action to the global situation of that moment, instead of relating actions among them. Representing user's behaviours by means of sequences, as well as creating complete and comprehensible representation, allows us to relate actions in terms of time. The group that has been working in MavHome and Casas projects [11] was one of the first groups that emphasized the importance of associating actions and their temporal relationships. They started using time intervals associated with activities rather than instantaneous time points. They used Allen's temporal logic [12] in order to represent time intervals, producing fairly intuitive sequences of actions. In that sense, considering Michael's example, the system would be able to detect that he first gets up, then he goes into the bathroom and then he turns on the lights. This system only considers Allen's temporal logic relations (which define relations qualitatively), ruling out quantitative relations, which allow us to define more accurate relationships. In that sense, Aztiria et al. [13] suggested a process to discover quantitative temporal relationships, but they only applied it to pairwise relationships.

4 Identifying Time Relations

The step of 'Identifying Topology' provides a sequential representation of patterns. This way of representation implies a first definition of temporal relationships due to the fact the sequence itself defines a temporal order of actions. In that sense, all relationships defined by the sequences are *qualitative* relationships defined by the term '*after*'.

Qualitative relationships allow us to understand the logic order of actions within a sequence. Even so, such relationships could be better defined if we were able to discover a regular *quantitative* relationship. Compared to *qualitative* relationships, *quantitative* relationships provide a higher quality information, being possible to use them for other purposes. One of those purposes can be the automation of devices, which is possible with *quantitative* relationships. We will consider Michael's behaviour of turning on the light of the bathroom 4 seconds after he goes into (defined by ActionPattern 0). If such a relationship is defined by means of a *qualitative* term like 'after', the system cannot infer when it has to turn on the light because it does not know if the time delay is 4 seconds, 5 minutes or 2 hours. However, *quantitative* relationships (4 seconds in Michael's case) allows the system to turn on the light at right time.

```
(ActionPattern 0)
  ON occurs (Bathroom, On,t0)
  IF context (Bathroom Light level is (<,10))
  THEN do (On, BahtLights, t) when t=t0+4s
```

Therefore, the objective of this step of 'Identifying Time Relations' is to discover *quantitative*, or accurate temporal relationships, to better define user's behaviours. For that, SPUBS includes two different algorithms. Following, these two algorithms as well as the process of getting necessary information will be described.

4.1 Data Collection for Identifying Time Relations

The first step to discover such relationships will be to collect necessary data. Relationships to study will already be defined by the topology. Thus, for each relationship defined by the topology we will collect particular time relationships of all occurrences.

Let us consider again Michael's example and the relationship between the actions 'Bathroom on' and 'BathLights on'. For each case where a occurrence of 'Bathroom on' is followed by a occurrence of 'BathLights on', SPUBS collects the time distance between both actions. Let us consider time relationships of Michael's example are depicted in Figure 2.

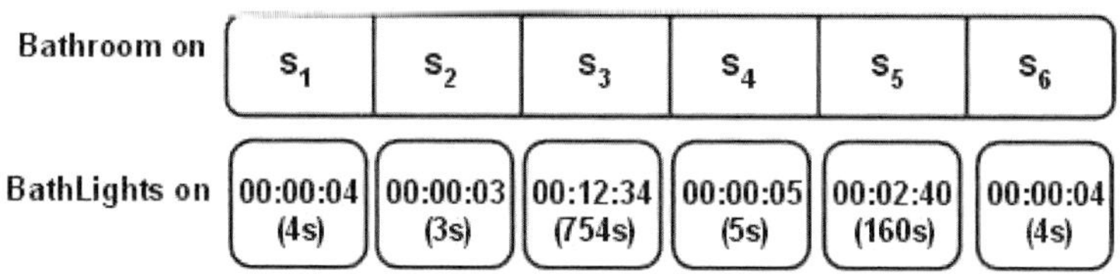

Fig. 2 Time Distances between occurrences of 'Bathroom on' and 'BathLights on'

Once temporal distances are collected, the next step is to identify interesting time relationships. For that, SPUBS includes two algorithms, 'Basic Algorithm' and 'EM Algorithm', so that the user of the system may choose any of them to identify such relationships. Both algorithms are based on the same idea of grouping occurrences taking into account their similarities and deciding if a group represents an interesting time relationship.

4.2 'Basic Algorithm' for Identifying Time Relations

The technique to make groups could be as complex as we can imagine and different techniques could be suggested to accomplish this task. SPUBS includes a very basic algorithm, which groups values that are within a range established by:

$$[min, max] = \overline{x} \pm (\overline{x} * tolerance) \qquad where \qquad \overline{x} = \frac{\sum_{i=1}^{n} a_i}{n} \tag{1}$$

with: tolerance = tolerated deviation from $\overline{x}$ *(%);* a_i *= time distance of an element; and n = number of elements.*

If a value does not fulfil the requirements to join any group, a new group is created using that value as group's mean value. Every time a new value is added to a group the mean value of that group is recalculated.

Considering Michael's example and time distances depicted in Figure 2, the process of making groups would be (let us consider 50% as tolerance percentage):

(e_1,4s); There is no group; create(group0, $\overline{x}$ (4s), [2,6])

Tolerance = 4 $\pm$ (4 * 0.5)

(e_2,3s); 3=[2,6]; join(group0, $\overline{x}$ (3.5s), [1.75,5.25])

Tolerance = 3.5 $\pm$ (3.5 * 0.5)

(e_3,754s); 754$\neq$[1.75,5.25]; create(group1, $\overline{x}$ (754s), [377,1132])

Tolerance = 754 $\pm$ (754 * 0.5)

(e_4,5s); 5=[1.75,5.25]; join(group0, $\overline{x}$ (4s), [2,6])

Tolerance = 4 $\pm$ (4 * 0.5)

(e_5,160s); 160$\neq$[2,6] and 160$\neq$[377,1132]; create(group2, $\overline{x}$ (160s), [80,240])

Tolerance = 160 $\pm$ (160 * 0.5)

(e_6,4s); 4=[2,6]; join(group0, $\overline{x}$ (4s), [2,6])

Tolerance = 4 $\pm$ (4 * 0.5)

Once groups are created, the following step is to decide what groups define interesting time relationships. For that, SPUBS's user must define a minimum confidence level. Thus, those groups that cover more instances than established by the minimum confidence level would be considered as interesting, being their mean value the temporal relationship. In Michael's case, let us consider a minimum confidence of 25%.Thus, the only group considered as interesting would be 'group 0', which covers 4 out of 6 occurrences.

4.3 *'EM Algorithm' for Identifying Time Relations*

Besides the 'Basic Algorithm', SPUBS includes another algorithm, which creates clusters of time distances based on the Expectation-Maximization (EM) algorithm [14]. The basic idea of the EM algorithm is to estimate the maximum likelihood between parameters. An important advantage of this algorithm is that it automatically calculates the necessary number of groups and includes each occurrence in its corresponding group.

Once groups are created by the EM algorithm, the process of deciding what groups defines an interesting time relationship is same as 'Basic Algorithm'.

5 Validation and Results

In order to validate the system we applied it to artificial data generated at the University of Ulster and then to a real dataset collected from Washington State University's (WSU's) Smart Apartment. The data collected in the WSU smart apartment [15] represents participants performing five Activities of Daily Living (ADLs) in the apartment: making a phone call, washing hands, cooking, taking medicine/eating the food and cleaning (See Table I for actions involved in each activity).

Table 1 Actions involved in each ADL

Activity	Involved Actions
Making a phone call	Phone Book (ON) –>Phone (ON) –>Phone (OFF)
Washing hands	Water (ON) –>Water (OFF)
Cooking	Cabinet (ON) –>Raisins (ON) –>Oatmeal (ON) –>Measuring spoon (ON) –>Bowl (ON) –>Sugar (ON) –>Cabinet (OFF) –>Water (ON) –>Water (OFF) –>Pot (ON) –>Burner (ON) –>Burner (OFF)
Taking medicine and Eating	Cabinet (ON) –>Medicine (ON) –>Cabinet (OFF) –>Water (ON) –>Water (OFF) –>Cabinet (ON) –>Medicine (OFF) –>Cabinet (OFF)
Cleaning	Water (ON) –>Water (OFF)

In all, the sensors installed in WSU smart apartment are:

- 14 sensors on objects such as phone, medicine container or cabinet.
- 27 motion sensors.

As the set of actions involved in these 5 ADLs and the order of such actions were known in advance, we knew what patterns should be discovered by SPUBS. As we expected, the steps of 'Identifying Frequent Sets' and 'Identifying Topology' discovered the actions involved in the sequence as well as their order. Once the topology was discovered we applied both algorithms, described in Section 4, to discover temporal relationships. The results obtained in that process showed that it was possible to discover an accurate temporal relationships in 11 out of 17 relationships. Following some of the ActionPatterns, i.e. parts of the sequence, discovered by the system are represented using $\mathscr{L}_{SPUBS}$:

```
(ActionPattern 1)
  ON occurs (Phone, On,t0)
  IF (by default = true)
  THEN do (Off, Phone, t)
       when t=t0+50s

(ActionPattern 2)
  ON occurs (Cabinet, On,t0)
  IF (by default = true)
  THEN do (On, Medicine, t)
       when t=t0+3s

(ActionPattern 3)
  ON occurs (Medicine, On,t0)
  IF (by default = true)
  THEN do (Off, Cabinet, t)
       when t=t0+2s

(ActionPattern 4)
  ON occurs (Cabinet, Off,t0)
  IF (by default = true)
  THEN do (On, Water, t)
       when t=t0+16s
```

6 Conclusions

Intelligent Environments need to know the common behaviours and preferences of the user in order to meaningfully act. Representing such common behaviours by means of sequences facilitates their understanding. Besides, such a representation allows us to relate such actions in terms of time. In that sense, SPUBS includes two algorithms to get more accurate relationships between actions. Accurate temporal relationships, represented by means of numerical values, defines better such relationships than using qualitative values, due to the fact they allow the system to use the sequences for additional purposes, for example to automate devices' activation.

References

1. Cook, D., Augusto, J., Jakkula, V.: Ambient intelligence: Technologies, applications, and opportunities. Pervasive and Mobile Computing 5(4), 277–298 (2009)
2. Nakashima, H., Aghajan, H., Augusto, J.C.: Handbook on Ambient Intelligence and Smart Environments. Springer, Heidelberg (2009)
3. Weiser, M.: The computer for the 21st century. Scientific American 265(3), 94–104 (1991)

4. Augusto, J.C., Cook, D.J.: Ambient intelligence: applications in society and opportunities for ai. In: 20th International Joint Conference on Artificial Intelligence, IJCAI 2007 (2007)
5. Galushka, M., Patterson, D., Rooney, N.: Temporal data mining for smart homes. In: Augusto, J.C., Nugent, C.D. (eds.) Designing Smart Home. The Role of Artificial Intelligence, pp. 85–108. Springer, Heidelberg (2006)
6. Aztiria, A., Izaguirre, A., Basagoiti, R., Augusto, J.C., Cook, D.J.: Discovering of frequent sets of actions in intelligent environments. In: Proceedings of the 5th International Conference on Intelligent Environments, pp. 153–160 (2009)
7. Aztiria, A., Izaguirre, A., Basagoiti, R., Augusto, J.C., Cook, D.J.: Automatic modeling of frequent user behaviours in intelligent environments. In: Proceedings of the 6th International Conference on Intelligent Environments (Submitted, 2010)
8. Cook, D.J., Das, S.K.: How smart are our environments? an updated look at the state of the art. In: Pervasive and Mobile Computing, vol. 3, pp. 53–73. Elsevier Science, Amsterdam (2007)
9. Hagras, H., Callaghan, V., Colley, M., Clarke, G., Pounds-Cornish, A., Duman, H.: Creating an ambient-intelligence environment using embedded agents. IEEE Intelligent Systems 19(6), 12–20 (2004)
10. Mozer, M.C., Dodier, R.H., Anderson, M., Vidmar, L., Cruickshank, R.F., Miller, D.: Current trends in connectionism. In: The neural network house: an overview, pp. 371–380. Erlbaum, Mahwah (1995)
11. Jakkula, V.R., Crandall, A.S., Cook, D.J.: Knowledge discovery in entity based smart environment resident data using temporal relation based data mining. In: 7th IEEE International Conference on DataMining, pp. 625–630 (2007)
12. Allen, J.: Towards a general theory of action and time. Artificial Intelligence 23, 123–154 (1984)
13. Aztiria, A., Augusto, J.C., Izaguirre, A., Cook, D.J.: Learning accurate temporal relations from user actions in intelligent environments. In: Proceedings of the 3rd Symposium of Ubiquitous Computing and Ambient Intelligence, vol. 51/2009, pp. 274–283 (2008)
14. Hogg, R., McKean, J., Craig, A.: In: Introduction to Mathematical Statistics, pp. 359–364. Pearson Prentice Hall, London (2005)
15. Cook, D., Schmitter-Edgecombe, M.: Activity profiling using pervasive sensing in smart homes. IEEE Transactions on Information Technology for Biomedicine (2008)

A Framework to Enable Two-Layer Inference for Ambient Intelligence

Xiang Su, Davide Fucci, and Jukka Riekki

In this paper, we present a two-layer inference framework to enable Semantic Web technology-based intelligent functionalities in ambient environments. The basic idea is that the low level inference is performed on the mobile devices capable of utilizing ontology, and only the high level inference is performed at the server side. This paper presents the design of this framework and illustrates its usability by a use case. The framework fully utilizes the computing capabilities of devices in the system and this way minimizes the communication among devices.

1 Introduction

Ambient environments will contain plentiful of devices with different capabilities in our surroundings and our lives. Some devices are resource-constrained, for example sensors in sensor networks have limited processing power, bandwidth, and memory. On the other hand, powerful mobile terminals can even contain knowledge-based systems. Therefore, it's a challenge to utilize the capabilities of devices fully to enable intelligent functionalities in the environment.

Semantic Web technologies offer good opportunities to tackle this challenge. Our aim is to utilize these technologies to build interoperable solutions for connecting the devices in the ambient environments and using the full capabilities of each device in offering a rich set of services for the users.

In this paper, we are proposing a framework to enable a two-layer inference according to the capabilities of different devices. In this framework, resource-constrained devices deliver measurement data to servers and mobile devices.

Xiang Su · Jukka Riekki
Intelligent Systems Group and Infotech Oulu, University of Oulu, Finland
e-mail: xiang.su@ee.oulu.fi, jukka.riekki@ee.oulu.fi

Davide Fucci
Dipartimento di Informatica, Università degli Studi di Bari, Italy
e-mail: fucci.davide@unibari.net

J.C. Augusto et al. (Eds.): ISAmI 2010, AISC 72, pp. 29–36.
springerlink.com

Ontology-enabled mobile devices perform simple inference tasks (low-level inference) based on the data from sensors. We define an ontology-enabled device as a device capable of utilizing ontology to model the application domain and to reason about the properties of the application domain. The server performs high-level inference, including all complex tasks like advanced inference, context extraction and reuse, coordinating the devices, etc.

This framework enables harnessing the computing capabilities of the system. The resource-rich servers support the Semantic Web based intelligent functionalities. On the other hand, some inference can be deployed to less capable devices like mobile terminals. For example, low level inference engine can be deployed on mobile devices to perform RDF [1] inference and simple OWL-based inference [2]. When inference is performed on ontology-enabled mobile devices there is no need to transfer sensory data, user profiles etc. to the server side. That is, the framework facilitates minimizing the amount of communication. This is an important prerequisite for the creation of a general framework for all the possible usages, because in many practical applications, capabilities of sensors and communication links in the system are limited.

The idea of a two-layer approach for inference has been reported by [3] and [4]. They addressed specific problems like building the semantic representation of the user profile and emotional status, but did not consider any general framework. We expect to develop a more general and extensible framework for supporting a wider set of devices and application scenarios. Besides, some general ambient reasoning solutions have been discussed before. For example, [5] proposed a context-aware architecture supporting context reasoning, [6] presented an intelligent system based on low level, distributed, fuzzy inference learning and [7] proposed a solution integrating reasoner into a middleware. Some of them only focused on high-level inference, while others focused on low-level. None of them considered the different capabilities of devices, not to mention communication. In our framework, we will consider harnessing the full capabilities ambient systems.

We will continue this article with introducing standards and technologies in section 2. Details of the design considerations of the framework will be presented in section 3, and a use case will be presented in section 4. Finally, we will draw conclusions in section 5.

2 Standards and Technologies

To enable a flexible, extensible, and effective intelligent framework, we are using standardized Semantic Web technologies, like RDF, OWL, and SWRL [8] rule-based inference engine. Besides, we are employing Entity Notation (EN), which is a lightweight data/knowledge representation for communication in the system innovated by [9]. It can support Semantic Web-based inference straightforwardly.

RDF (Resource Description Framework) is currently the standard for establishing semantic interoperable systems. It facilitates interoperation because it provides a data model that can be extended to address ontology representation, and it enables the automatic processing for resources. In our framework, RDF is utilized as a low level representation for expressing knowledge. For example, sensory data will be transformed into RDF statements as the first step for the semantic processing. We encode RDF triplets using RDF/XML, which has been proposed by [10] as a standard to encode RDF models.

To represent the ontology layer knowledge, the W3C suggests OWL as a standard [11]. OWL ontology consists of a set of axioms which place constraints on sets of individuals and the types of relationships permitted between them. These axioms provide semantics by allowing systems to infer additional knowledge based on the data provided explicitly. In this framework, we are using OWL as the main representation for context modelling as the basis for the inference. Both high-level and low-level inferences need ontology in the reasoner.

Rule-based inference is utilized to derive new knowledge from the assertions in ontology at the server side. We adopt SWRL rules; a rule presents an implication between an antecedent and a consequent so that the intended meaning is: whenever the conditions specified in the antecedent hold, then the conditions specified in the consequent must also hold. The atoms composing both parts can be expressed in the form C(x), P(x,y), sameAs(x,y) or differentFrom(x,y), where C is an OWL description, P is an OWL property, and x,y are either variables, OWL individuals or OWL data values.

Entity Notation is proposed as a lightweight representation for transforming information between resource-constrained devices and knowledge-based intelligent systems. This representation can be handled by resource-constrained devices, transferred over modest communication links, and transformed into knowledge (RDF and ontology) in a straightforward fashion. It's a practical data representation with expressive power. We utilize EN as the main representation for data and knowledge exchange in this framework.

3 Designing the Framework

The framework we are proposing depicts a two-layer structure that enables different inference functionalities for different devices. Figure 1 shows a concept diagram of our framework, which includes an ontology (consisting of T-Box and A-Box) and a rule set. The T-Box includes assertions on concepts, while the A-Box includes assertions on individuals, which are RDF statements. In the T-Box, all concepts and relations from the environment are described as one part of context, even though we do not give an overall concept hierarchy in this picture.

EN facilitates the transformation from different data sources (sensor data, user profiles, etc.) to RDF statements, and RDF statements can be mapped to the concepts in the T-Box unambiguously. The low-level inference makes use of A-Box level and small parts from T-Box level. The usage of T-Box level knowledge enables low level inference to have stronger capability than simple RDF inference, because of the description logic inference from the T-Box. But the low level cannot support any designer-specific rule-based inference. High level inference engine can support

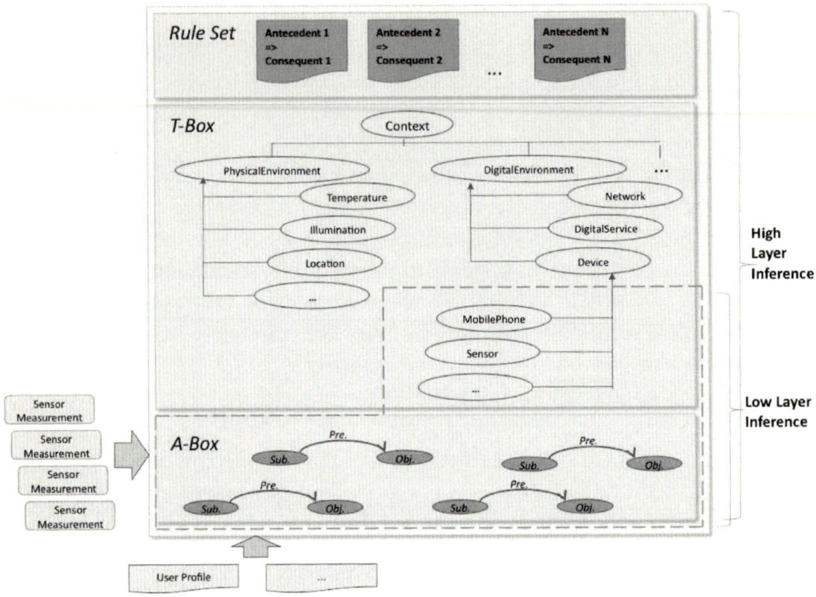

Fig. 1 Concept Diagram of Two-Layer Inference Framework

the full capabilities of ontological inference and rule-based inference. Complex intelligent functionalities, like multiple user interactions and device interoperability, can be realized at high level inference.

Figure 2 illustrates how the conceptual diagram can be deployed at devices in ambient environments. A server, ontology-enabled mobile devices, and resource-constrained sensor nodes form of a two-layer inference system. Knowledge-based systems manage domain ontology at the server side, which host all advanced knowledge-based applications. All Semantic Web functionalities, like reasoning, publishing semantic data and semantic matching, can be implemented at the server side. Mobile devices share parts of domain ontology from the server and perform low-level inference based on shared ontology and RDF statements. Some other devices, like sensors, have really limited resources. They can send data to the nearest intelligent mobile devices. The data can then be forwarded to the server, when necessary. Servers and mobile devices can send commands to some sensors as well. The data flow of a specific system can be decided at the design time. Data is delivered as EN packets. Generally, sensor data for realizing self-adaptive functions of devices can be sent to ontology-enabled devices, while sensor data for realizing complex functions, like multi-user interactions and rule-based inference, should be sent to a server.

Entity Notation plays an important role in the communication of this framework. A knowledge-based system on a server can compose EN packets to transfer parts of the ontology to ontology-enabled mobile devices. Then these devices decompose EN packets and get the knowledge they need in their intelligent applications.

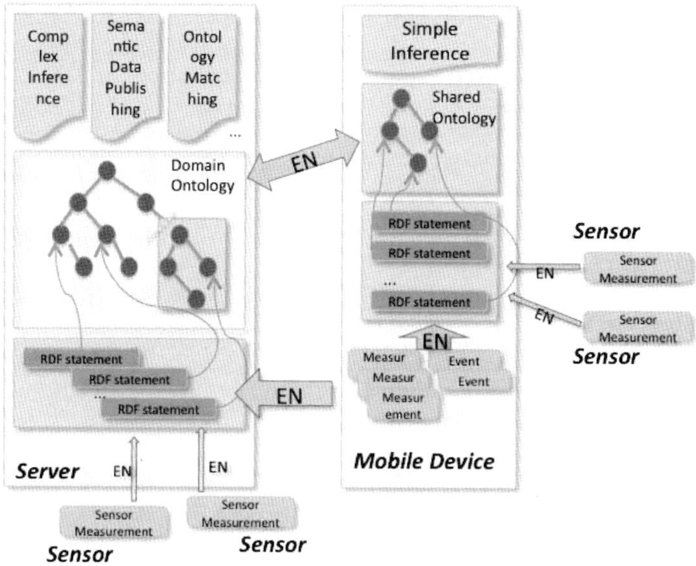

Fig. 2 Framework of two-layer inference in ambient environments

Ontology-enabled mobile devices can also use EN composer to compose EN packets and transfer them to knowledge-based systems on servers. Resource-constrained devices cannot handle any ontology or reasoner. But still, they can compose EN packets based on their measurements and transfer them to ontology-enabled mobile devices or servers. Knowledge-based systems can reason on the transferred data and offer semantic support for these resource-constrained devices. More details about EN can be found from the paper [9].

4 A Use Case

In this section we are presenting a use case to illustrate the usability of our framework. This example demonstrates how data is passed among main components, and how simple inference and complex inference is achieved. A social application scenario is as follows.

A couple is driving home from different working places after working, and one of them needs to pick up their children from the kindergarten. The couple has decided to pick up their children as early as possible. When a parent has left the workplace she/he is reminded to pick up the children if she/he is the nearest parent already left the work. At the same time, another reminder will be shown to suggest the spouse to drive home directly.

Devices to enable the two-layer inference for this scenario include: simple GPS receivers for children, a server and ontology-enabled mobile devices with RFID readers, GPS receivers and map applications for parents. When a parent gets off work, he touches with his phone an RFID tag nearby the entrance of the building. An EN packet like this records the time he gets off:

```
[urn:uuid:739ae5 "EntranceB"
    "JohnSmith" "2010-03-25T16:21:40"]
```

In this packet, urn:uuid: 739ae5 determines the template for this packet and facilitates the transformation from an EN packet to a RDF statement. Here is the corresponding RDF statement of this EN packet. We utilize xmlns:e="http://ee.oulu.fi/o#" in this statement.

```
<e:RfidTag rdf:ID="rfid21">
   <e:tagType rdf:datatype=
  "http://www.w3.org/2001/XMLSchema#String">
  ExitPlace</e:tagType>
  <e:personTouching rdf:datatype=
  "http://www.w3.org/2001/XMLSchema#String">
  JohnSmith</e:personTouching>
  <e:timeTouched rdf:datatype=
  "http://www.w3.org/2001/XMLSchema#dateTime">
  2010-03-25T16:21:40</e:timeTouched>
</e:RfidTag>
```

Low level inference can be performed based on this RDF statement and an ontology. In the following ontology snippet, a person who leaves the building after 3 p.m. is assumed to go home. The embedded reasoner on ontology-enabled device deduces that John Smith is going home, and starts to send GPS data to the server.

```
<owl:Class rdf:ID="PersonGoingHome">
  <rdfs:subClassOf rdf:resource="#Person" />
  <rdfs:subClassOf>
    <owl:Restriction>
      <owl:onProperty rdf:resource="#timeTouched"/>
      <owl:mintimeCardinality rdf:datatype="&xsd;time">
      15:00:00</owl:cardinality>
    </owl:Restriction>
  </rdfs:subClassOf>
</owl:Class>
```

GPS packet with user's name, longitude and latitude data is like this in EN format:

```
[urn:uuid:7bcf39 "JohnSmith" "25.47" "65.06"]
```

This packet can also be transformed into a RDF statement. In this scenario, high level inference can be applied for interaction among multiple devices. For example, mobile map applications on the device can estimate the driving distance of the parents. And high level inference can decide who should visit the kindergarten by using a following SWRL rule:

```
Implies(Antecedent drivingdistance(?person1, ?dis1)
                   drivingdistance(?person2, ?dis2)
                   swrlb:greaterThan(?dis1, ?dis2)
Consequent(showReminder(?person1 "Please visit kindergarten
to pick up your kids")
           showReminder(?person2 "Please go home directly"))
```

The above example shows the utilization of the inference engine at two levels, which enable reasoning of different functionalities. More complex reasoning can be performed based on the data from more sensors and users in a real implementation.

5 Discussion

We presented a novel framework to enable two-layer inference in ambient environments. This framework can utilize the computation capability of the system at its best, while minimizing the communications. We introduced the requirements, technologies, and design considerations in details. A use case has been presented to illustrate the usefulness of our framework.

Three kinds of devices with different capabilities had been considered: resource-constrained sensors, ontology-enabled mobile terminals and servers. Two-layer inference framework can be deployed on ontology-enabled terminals and servers respectively. Ontology-enabled terminals support fast and local reasoning to react the changing environment, while servers afford complex inference and other resource-consuming semantic functionalities. The proposed framework is a scalable solution as the utilization of unique identifiers (Universally Unique Identifiers and URIs) in EN packets.

As future work, we will verify this framework with a real world implementation, which includes the server, mobile devices and very small sensors that can send measurements in the EN format. Moreover, we will study the structure of the domain ontology to facilitate the sharing of knowledge in a plug-in fashion. Finally, we will consider the optimization of the inference engine to minimize the computing resources for mobile devices.

Acknowledgements. We thank Dr. Jouni Markkula and Ekaterina Gilman for their advice for this paper. The first author would thank the funding from the Infotech Oulu Graduate School and the MOTIVE research program of the Academy of Finland.

References

1. Carroll, J.J., Klyne, G.: Resource description framework (RDF): Concepts and abstract syntax. In: W3C recommendation, W3C, Feburary (2004), http://www.w3.org/TR/2004/REC-rdf-concepts-20040210/
2. McGuinness, D.L., van Harmelen, F.: OWL: Web Ontology Language Overview. In: W3C recommendation, W3C (February 2004), http://www.w3.org/TR/2004/REC-owl-features-20040210/

3. Agostini, A., Bettini, C., Riboni, D.: Loosely coupling ontological reasoning with an efficient middleware for context-awareness. In: Proceedings of the 2nd Annual International Conference on Mobile and Ubiquitous Systems: Networking and Services (MobiQuitous 2005), pp. 175–182. IEEE Computer Society, Los Alamitos (2005)
4. Acampora, G., Loia, V., Nappi, M., Ricciardi, S.: Hybrid computational intelligence for ambient intelligent environments. In: Atlantic Web Intelligence Conference, pp. 26–31 (2005)
5. Goh, E., Chieng, D., Mustapha, A.K., Ngeow, Y.C., Low, H.K.: A context-aware architecture for smart space environment. In: Multimedia and Ubiquitous Engineering, pp. 908–913 (2007)
6. Wang, K., Abdulla, W., Salcic, Z.: Multi-agent fuzzy inference control system for intelligent environments using Jade. In: 2nd IET International Conference on Intelligent Environments (IE 2006), July, vol. 1, pp. 285–294 (2006)
7. Xu, C., Cheung, S.-C., Lo, C., Leung, K.C., Wei, J.: Cabot: On the ontology for the middleware support of context-aware pervasive applications. In: Network and Parallel Computing, pp. 568–575 (2004)
8. Horrocks, I., Patel-Schneider, P.F., Boley, H., Tabet, S., Grosof, B., Dean, M.: SWRL: A semantic web rule language combining OWL and RuleML. In: W3C recommendation, W3C (May 2004), http://www.w3.org/Submission/2004/SUBM-SWRL-20040521/
9. Riekki, J., Su, X., Haverinen, J.: Connecting resource-constrained robots to knowledge-based systems. In: Proceedings of International Conference on Modelling, Identification, and Control (MIC 2008), ACTA Press (2008)
10. Beckett, D.: RDF/xml syntax specification (revised). In: W3C recommendation, W3C, Feburary 2004, Feburary (2004), http://www.w3.org/TR/2004/REC-rdf-syntax-grammar-20040210/
11. Herman, I.: Overview of the W3C technologies. tech. rep., W3C (2006), http://www.w3.org/Consortium/Offices/Presentations/Overview/6

GLSS – Group Learning in Shared Spaces Considering Aspects Like Emotion and Personality

Goreti Marreiros, Ricardo Santos, and Carlos Ramos

Abstract. Most students experience learning difficulties in various points during their school years. When new contents are presented the reactions can be different depending on the individual characteristics. In this paper we present an architecture that intends to combine agent technology with computational models of personality and emotion, within an ubiquitous learning system that is able to support the learning process among a group of students coordinated by an instructor.

1 Introduction

Education is one of the greatest assets that people can acquire. However students can have some difficulties associated with the learning process like for instance. confidence, students who feel they can do well will try even if the work is difficult; emotional control, students able to control their emotions do not panic if test questions are hard; motivation, usually students feel more motivated when their progress is emphasized by pointing out that what they are doing now is more complicated than the work of six weeks ago. The instructors are another important component in the learning process, many of the problems pointed previously can be addressed by them but usually the instructors do not have real time access to the student's problems and overall performance and when they have its too late.

Goreti Marreiros · Ricardo Santos · Carlos Ramos
GECAD – Knowledge Engineering and Decision Support Group Porto, Portugal
e-mail: {goreti,csr}@dei.isep.ipp.pt

Goreti Marreiros · Carlos Ramos
Institute of Engineering – Polytechnic of Porto, Porto, Portugal

Ricardo Santos
School of Management and Technology of Felgueiras – Polytechnic of Porto
Felgueiras, Portugal
e-mail: rjs@estgf.ipp.pt

J.C. Augusto et al. (Eds.): ISAmI 2010, AISC 72, pp. 37–44.

Student's strategies for processing information in classroom may include aspects like: seeing, hearing, reflecting and acting, reasoning. Another variable that affects the comprehension of new information are the teaching strategies. Some teachers just discourse, others demonstrate and/or guide students to self-discovery, some focus on theory and others in application, some highlight memory and others understanding. If there is a divergence between students learning styles and professor teaching style, this may result in bored and inattentive students. On the other hand in an ideal perspective the professor teaching style should be adapted to the student individual learning style and student progression.

In this paper we present our vision that a truly effective intelligent tutoring system would need to embody the understanding of the various mechanisms underlying human personality, emotions, and intentions and how these influence the learning process of a student. Our aim is to incorporate these processes within an intelligent tutoring system by leveraging on relevant principles from both the socio-cognitive and psychological domains. GLSS (Group Learning in Shared Spaces considering aspects like emotion and personality) aims to combine agent technology with computational models of personality and emotion, within an ubiquitous learning system that is able to support the learning process among a group of students coordinated by an instructor.

2 Background

In this section we will perform a brief description of the main areas covered by GLSS project: learning styles, intelligent tutoring, emotion and personality.

2.1 Learning Styles

Students have different levels of motivation and different attitudes about teaching and learning. This will influence students' response to specific classroom environments and instructional practices. In this context, the more thoroughly instructors understand these differences, the better chance they have of meeting the diverse needs of their learners [1].

There are three main categories or philosophical frameworks under which is possible to classify learning theories, namely: behaviorism, cognitivism, and constructivism. According to [2] constructivism is the dominant learning theory of the last decade. In this theory knowledge is actively constructed by the student with an adaptive process. Students do not analyze the information in a static and hermetic way but search for blocks of related old knowledge to construct a new and more significant knowledge.

Learning activities in constructive perspective involve [3]:

- a positive motivational context, hopefully intrinsic but at least one involving a felt need-to-know and a aware emotional climate;
- a high degree of learner activity, both task-related and reflective;
- interaction with others, both at the peer level with other students, and hierarchically, within "scaffolding" provided by an expert tutor;

- a well-structured knowledge base, that provides the longitude or depth for conceptual development and the breadth, for conceptual enrichment.

The main point of constructivism theory is that students should be consciously involved in the learning process. According to this theory the role of teachers is to guide the students [4] in the construction of new knowledge, which is generally based in student's individual background. The emphasis in student individual differences is really important to recognize, design and support students activities. In constructivism learning theory, students have different learning styles. Also, the capacity of adaptation in different social contexts and the constructive social aspect of knowledge must be taken in consideration [4].

Generally, learning styles is understood as something that intent to define models of how determinate person learns. Generally it is understood that each person has a learning style different and preferred with the objective to achieve better results.

Some case studies have been proposed that teachers should assess the learning styles of their students and adapt their classroom and methods to better fit each student's learning style [5].

2.2 Intelligent Tutoring

Intelligent tutoring systems (ITS) appeared during the 1970s, mostly driven by the success of knowledge-based systems and expert systems. ITS' are able to instruct and train students and professionals without the intervention of human beings [18-19].

Developments were made in trainees' models, instructional and pedagogical planning, and user interface. In the 1990s, with the Web boom, some ITS ideas were incorporated in new computer-aided instruction paradigms, like e-learning and distributed learning.

However, there is a clear difference in the level of interactions and types of skills addressed by ITS and other e-learning systems. In the last several years, ITS have moved from research labs into the real world [20]. Many systems were developed and deployed, even for critical and complex domains. The reported benefits demonstrated by the users of these systems are impressive. ITS-taught trainees generally learn faster and translate the learning into improved performance better than classroom-trained participants.

Today, ITS can be produced by authoring tools [21-22], and specific evaluation and assessment methods can be used [23], appearing sometimes in the form of Cognitive Tutors [24].

2.3 Emotion

The psychology literature is replete with examples of how emotions affect the individual learning process [6,7]. The emotions have proven affect on the school activities related to attention and the emotional system besides being complex does not prone to change [8]. The frequently changing emotional states of an individual influence their behavior and their interactions with those around them—in the present context, other group members. For example, the phenomenon of emotional contagion is the tendency to express and feel emotions that are similar to and influenced by those of others. This phenomenon could be analyzed the modal mood of the group in terms of a particularly salient emotion that one of the group members is feeling [9,10].

We propose to incorporate emotion into our system using the OCC model of Ortony, Clore and Collins [11], a model that is widely used for emotion simulation of embodied agents [12-15], already used in past projects by our own group [16]. In the OCC model, agent's concerns in an environment are divided into goals (desired states of the world), standards (ideas about how people should act) and preferences (likes and dislikes), and distributed across twenty-two formally representable emotion categories or "types". To reduce the complexity in his original model, Ortony proposed a new simplified model with 10 emotional categories divided in five positive (joy, hope, relief, pride, gratitude, and love) and five negative categories (distress, fear, disappointment, remorse, anger, and hate) [17]. We expect this reduced model to be adequate for our purposes.

2.4 Personality

Personality differences manifest themselves in differences in all aspects of psychological life--in affect, behavior, motivation, perception and cognition. Whereas there are disagreements amongst personality psychologists as to the best way to conceptualize these individual differences, there is considerable evidence in support of what is called the Five Factor Model [26]. According to this model, which is the most widely model of personality used in computer applications, individual differences can be captured in terms of the five trait dimensions of Openness, Conscientiousness, Extraversion, Agreeableness and Neuroticism [27]. Like emotion, personality as also impact on the learning process [28].

3 Proposed Architecture

3.1 Architecture

Our main idea is to design a ubiquitous agent-based architecture, which will support geographically-distributed students. This would model (through intelligent agents) the personalized coaches for each student that thought intelligent shared spaces can have access to the topics proposed by the instructor and supported by the coaches. In figure 1 it is possible to see the proposed architecture.

The GLSS system is composed by two main modules: instruction and learning. The instruction module will be related to the teacher's actions and feedback from the students. The learning module will be focus on the student learning process. A

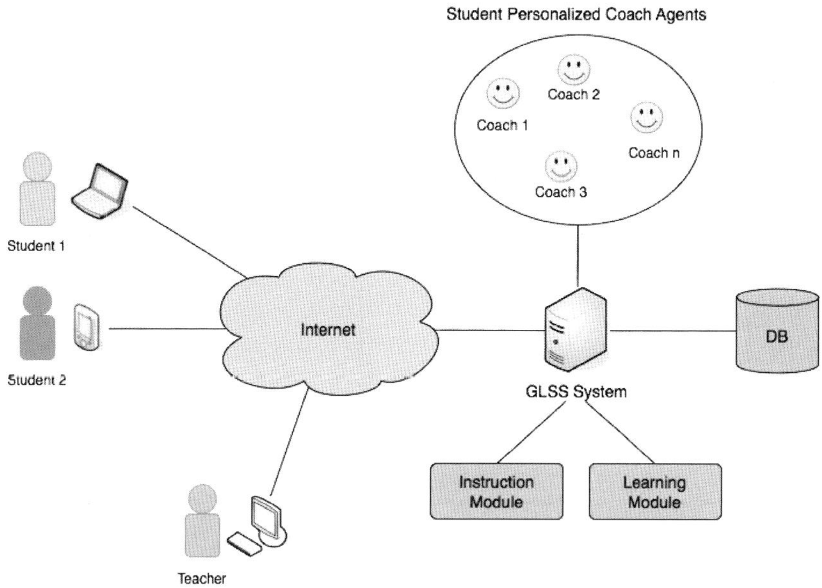

Fig. 1 GLSS Architecture

database will be needed to store all the topics, exercises, etc proposed by the teacher as well as the evolution of the students. To allow the ubiquity of the students between shared spaces it will be used agent-based modeling with web services. This way, their coaches will support students in all stages, and it will also be possible to receive the topics and feedback from the instructors. To overcome geographical and time constraints such that the agents (who represent the student coach) can report the student situation and receive new instructions for future sessions the system will be persistent in the server side. The received instructions from the instructors are used by the students' coach planning when the student is not available. Another action that the coach can do is the analysis of the learning process in order to understand if the technique applied is the adequate or if it is necessary to change the learning strategy.

The main role in this process will be played by the relation between the students and the personalized coach agents. The coach agents are built on an architecture that incorporates aspects like personality, emotions and intention in order to create a student computational profile that it is monitored. The emotion and the personality of the students is considered in the leaning process, the emotion is important in the evaluation when the student succeeds or fails the exercises and the personality is more important in the relation with the coach. Continual interaction with the student allows the agent to gather additional facts or make inferences about the student, e.g. motivation level, learning style, personality, level of interest, performance, level of patience, attention span, and so on. Agents will tailor experience to match the child's characteristics. E.g. being more encouraging for shy or introverted students who may know answers to questions. Or offer clues to weaker students on how certain problems could be solved, reminding them of previous problems they had solved so that they may recall the methods used, and so on. This reinforces their learning. For stronger students, having to put up with the progress of the weaker students may cause them to lose interest or patience or concentration. Instead, knowing so, the agents may, offer them additional or more advanced material that is more challenging, and keep them engaged.

3.2 Scenario

As the system is intended to support students learning in a ubiquitous manner, it would be tested in scenarios in which participants are distributed geographically or across time. Furthermore, we expect to test this system with students from diverse backgrounds under the influence of different personalities, emotions, and user intentions can emerge.

A possible scenario would be the following:

- Child is introverted. Might be too shy to answer questions even if child might know the answer. Possibly lacks confidence, etc.
- Agent knows this
- Agent encourages the child to answer questions raised by the teacher / organizer
- Especially if child has been performing well based on interactions with the agent (in various learning exercises on the same topic),
- or given the progress monitoring performed by the agent
- Hence the agent might encourage the child to answer questions or
- Agent may inform teacher that this child might know the answer
- Thus, child receives opportunity to answer questions
- Teacher would notice the child too
- Child would be encouraged
- Child may be happier
- Encourage equal participation for all children / students

4 Conclusions

This work proposes an architecture to support a system for group learning in shared spaces. Although the system provides teachers and students support, it is more focus in the students learning process. It considers aspects like emotion, and personality of the students to define the best strategy for the learning process supported by a personalized coach. The ubiquity of the system was defined as a prerequisite in early studies because students tend to have different spaces where they like to study most, individually or in group.

References

[1] Felder, R.M., Brent, R.: Understanding Student Differences. Journal of Engineering Education 94(1), 57–72 (2005)
[2] Karagiorgi, Y., Symeou, L.: Translating Constructivism into Instructional Design: Potential and Limitations. Educational Technology & Society 8(1), 17–27 (2005)
[3] Biggs, J.B.: Approaches to the enhancement of tertiary teaching. Higher Education research and Development 8, 7–25 (1989)
[4] Jonassen, D.: Objectivism versus constructivism: Do we need a new philosophical paradigm? Educational Technology, Research and Development 39(3), 5–14 (1991)
[5] Kolb, A., Kolb, D.A.: Learning styles and learning spaces: Enhancing experiential learning in higher education. Academy of Management Learning and Education 4(2), 193–212 (2005)
[6] Goleman, D.: Emotional intelligence: why it can matter more than IQ. Bantam Books, New York (1995)
[7] Kort, B., Reilly, R., Picard, R.W.: An Affective modle of interplay between emotions and learning: reengineering educational pedagogy- building a learning companion. In: Proceedings of IEEE International Conference on Advanced Learning Tehcnologies, Madison (2001)
[8] Sylwester, R.: How emotions affect learning. Educational Leadership, 60–65 (October 1994)
[9] Barsade, S.: The Ripple Effect: Emotional Contagion and Its Influence on Group Behavior. Administrative Science Quarterly 47, 644–675 (2002)
[10] Neumann, R., Strack, F.: Mood contagion: The automatic transfer of mood between persons. Journal of Personality and Social Psychology 79, 211–223 (2000)
[11] Ortony, A., Clore, G.L., Collins, A.: The cognitive structure of emotions. Cambridge University Press, Cambridge (1988)
[12] Gratch, J., Marsclla, S.: Evaluating a computational modcl of cmotion. Journal of Autonomous Agents and Multiagent Systems 11(1), 23–43 (2006)
[13] Mourão, D., Paiva, A.: EToy: Building an affective physical interface. In: Proceedings of the 2nd Workshop on Attitude, Personality and Emotions in User-Adapted Interaction (2001)
[14] MeiYii, L., Dias, J., Aylett, R., Paiva, A.: Improving Adaptiveness in Autonomous Characters. In: Prendinger, H., Lester, J.C., Ishizuka, M. (eds.) IVA 2008. LNCS (LNAI), vol. 5208, pp. 348–355. Springer, Heidelberg (2008)
[15] Bída, M., Brom, C.: Towards a platform for the education in emotion modeling based on virtual environments. In: Reichardt, D. (ed.) Proceedings of the 3rd Workshop on Emotion and Computing: Current Research and Future Impact, Kaiserslautern, Germany, September 23, pp. 45–52 (2008)
[16] Marreiros, G., Santos, R., Ramos, C., Neves, J.: Context Aware Emotional Model for Group Decision Making. IEEE Intelligent System (2010) (accepted for publication)
[17] Ortony, A.: On making believable emotional agents believable. In: Trapple, R.P. (ed.) Emotions in humans and artefacts, MIT Press, Cambridge (2003)
[18] Carbonell: AI in CA!: An artificial intelligence approach to computer-assisted instruction. IEEE Transactions on Man-Machine Systems 11, 190–202 (1970)
[19] Wenger, E.: Artificial intelligence and tutoring systems: computational and cognitive approaches to the communication of knowledge. Morgan Kaufmann Publishers Inc., San Francisco (1987)

[20] Ramos, C., Frasson, C., Ramachandran, S.: Introduction to the Special Issue on Real World Applications of Intelligent Tutoring Systems. IEEE Transactions on Learning Technologies 2(2), 62–63 (2009)

[21] Brusilovsky, P.: Developing adaptive educational hypermedia systems: From design models to authoring tools. In: Murray, T., Blessing, S., Ainsworth, S. (eds.) Authoring Tools for Advanced Technology Learning Environment, pp. 377–409. Kluwer Academic Publishers, Dordrecht (2003)

[22] Heffernan, N.T., Turner, T.E., Lourenco, A.L.N., Macasek, M.A., Nuzzo-Jones, G., Koedinger, K.R.: The ASSISTment Builder: Towards an Analysis of Cost Effectiveness of ITS creation. In: Proceedings of the 19th International Florida Artificial Intelligence Research Society Conference, pp. 515–520 (2006)

[23] Beck, J., Chang, K.-m., Mostow, J., Corbett, A.: Does help help? Introducing the Bayesian Evaluation and Assessment methodology. In: 9th International Conference on Intelligent Tutoring Systems, pp. 383–394 (2008)

[24] Aleven, V., Koedinger, K.R.: An effective meta-cognitive strategy: learning by doing and explaining with a computer-based Cognitive Tutor. Cognitive Science 26(2), 147–179 (2002)

[25] Marreiros, G., Machado, J., Ramos, C., Neves, J.: Argumentation-based Decision-Making in Ambient Intelligence Environments. International Journal of Reasoning-based Intelligent Systems (IJRIS) 1(3/4), 182–190 (2009)

[26] McCrae, R.R., John, P.O.: An introduction to the five-factor model and its applications. Journal of Personality 60, 175–215 (1992)

[27] Howard, P.J., Howard, J.M.: The BIG FIVE Quickstart: An introduction to the five-factor model of personality for human resource professionals. Center for Applied Cognitive Studies, Charlotte (2004)

[28] De Raad, B., Schouwenburg, H.C.: Personality in learning and education: a review. European Journal of Personality 10, 303–336 (1996)

Rewiring Strategies for Changing Environments

Wim Laurier, Geert Vanderhulst, Geert Poels, and Kris Luyten

Abstract. A typical pervasive application executes in a changing environment: people, computing resources, software services and network connections come and go continuously. A robust pervasive application needs adapt to this changing context as long as there is an appropriate rewiring strategy that guarantees correct behavior. We combine the MERODE modeling methodology with the ReWiRe framework for creating interactive pervasive applications that can cope with changing environments. The core of our approach is a consistent environment model, which is essential to create (re)configurable context-aware pervasive applications. We aggregate different ontologies that provide the required semantics to describe almost any target environment. We present a case study that shows a interactive pervasive application for media access that incorporates parental control on media content and can migrate between devices. The application builds upon models of the run-time environment represented as system states for dedicated rewiring strategies.

Keywords: Dynamic Pervasive Environments, MERODE, ReWiRe, Parental Control.

1 Introduction

Mobile devices such as a smart phone or ultra-mobile PC (UMPC) gain popularity and move towards interoperability with pervasive environments. To enable seamless integration of interoperable devices in new environments and to deal with purpose changes (e.g. the role of a mobile phone can evolve from multimedia

Wim Laurier · Geert Poels
Department of Management Information and Operational Management,
Faculty of Economics and Business Administration, Ghent University,
Tweekerkenstraat 2, 9000 Ghent, Belgium
e-mail: {wim.laurier,geert.poels}@ugent.be

Geert Vanderhulst · Kris Luyten
Hasselt University – transnationale Universiteit Limburg – IBBT Expertise Centre for
Digital Media, Wetenschapspark 2, 3590 Diepenbeek, Belgium
e-mail: {geert.vanderhulst,kris.luyten}@uhasselt.be

J.C. Augusto et al. (Eds.): ISAmI 2010, AISC 72, pp. 45–53.

device to VoIP device), there is a growing need for reconfigurable software applications that can dynamically adapt to their runtime environment [1, 2]. Therefore, this paper presents a combination of ontology-based environment models, which are constructed using the MERODE methodology [3-8] and ReWiRe's environment model [9], which enables software services to react to environment changes (e.g. by. the redistribution of a user interface to another device). Consequently, we rely on the MERODE methodology to create environment models that go beyond what is needed for creating pervasive environments. The MERODE framework has a strong theoretical underpinning in process algebra and supports dealing with asynchronous and parallel events by which a pervasive environment is characterized. This combination of reconfigurable environment models and pervasive software architecture (i.e. the ReWiRe's technological component is based on OSGi technology[1]) ensures reliable support for building context-aware applications [10] that can be rewired[2] at runtime.

Section 2 introduces the envisioned scenario, where section 3 shows the environment model that is used by the prototype application and section 4 illustrates the prototype application. Conclusions, limitations and directions for future research are discussed in section 5.

2 Scenario: Parental Control on Media Content in Dynamic Pervasive Environment

In our envisioned scenario, different types of media are shared on personal devices such as a mobile phone or set-top box with built-in hard disk. When brought together in a connected environment (i.e. a home network), shared media are discovered and listed in a user interface on end-user devices from where they can be accessed and streamed to an output device of choice. For example, an output device could be the current interaction device or a television set capable of playing media (i.e. running a media service). To protect children from content not suited for them (i.e. adult content) parental control is required to successfully deploy a pervasive media application in a real-world environment. Since the environment is assembled at runtime from available devices and media resources, parents have no absolute control over the media resources presented to their children. Therefore, the parental control has to be integrated into the environment model.

This usage scenario demands for a dynamic application that can adapt at runtime to changes in the environment configuration, as changes in the environment configuration can have an impact on the application's execution flow. Consequently, programming such an application in an ad-hoc way would be cumbersome due to the lack of a dynamic knowledge base reflecting the current state of the environment. The impact of environment changes can be indicated considering the example of a television streaming media residing on a mobile phone. If the phone and its owner leave the environment, the media stream is likely to be interrupted and the

---

[1] `http://www.osgi.org/`

[2] Rewiring is the dynamic reconfiguration process that enables systems to adapt themselves when their context of use changes.

application will need to cleanup allocated resources (e.g. shutdown a media service and switch off the television screen). Alternatively, when an output device is about to become unavailable, the application might react by automatically selecting another suitable output device to play the current media stream on.

3 Conceptual Models for Parental Control Application

We use the MERODE methodology to create a consistent[3] set of models, which reflect the current environment, and maintain consistency during model changes, which reflect environment changes. A key feature of the MERODE methodology is the use of existence dependency diagrams which relate the existence of a class instance with the existence of one of the instances in its parent class. Such dependency relations mean that instances of the existence dependent class refer to one and the same instance of the parent class during their entire lifespan. Fig. 1 shows the existence dependency diagram for our parental control application. The diagram is represented as a UML class diagram, in which the keywords on top of the class name indicate the origin of the concept that is represented by the class (the RESOURCE class originates in the ReWiRe ontology [9]). Existence dependency relationships are represented as dependency arrows of which the arrowhead

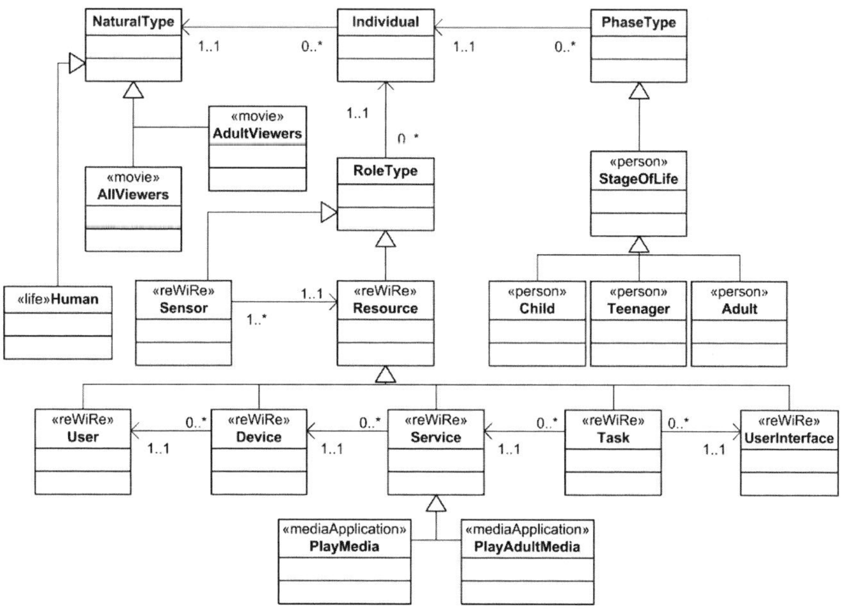

Fig. 1 Conceptual Model of Integrated Ontologies for Parental Control Application

[3] The consistency should prevent children from gaining access to content that is not suitable for them due to weaknesses in the environment model.

points towards the parent class (e.g. NATURALTYPE is a parent class for INDIVIDUAL). The existence dependency semantics are further specified by cardinality constraints added to these arrows. To prevent cognitive overload, methods and attributes were omitted and the 'movie', 'life' and 'person' ontologies are purpose-built minimal ontologies and not fully-fledged ontologies.

The partial ordering of objects, which originates in the existence dependency semantics, determines the object event table, which is a second model (next to the existence dependency diagrams) used by the MERODE methodology and specifies the objects that are affected (i.e. class instances whose state is changed) by the events listed. The matching of objects and events is done at the type level such that the effect that events have on object states can be specified as class methods. MERODE's *event propagation* rule states that the methods that apply to an existence dependent class also apply to its parent class(es), where they can exhibit a different behaviour (i.e. polymorphism). Consequently, an event that creates an object, ends the life of an object or modifies the state of an object may also modify the state of the object's parent object or other objects further up the existence dependency chain. Apart from the existence dependency diagram and the object event table, for each class a lifecycle is specified that describes all possible sequences of object state changes caused by events. These lifecycle models build, together with the existence dependency diagram and the object event table, a conceptual schema for the application.

The conceptual backbone of the reconfigurable conceptual environment model, originates in the work of Parsons and Li [11], who distinguish natural, phase and role types. *Natural types* (e.g. human) are independent (i.e. the existence of their instances does not depend upon other instances) and are rigid (i.e. their individuals cannot migrate to another natural type during their lifetime). *Phase types* (e.g. child, teenager, adult) are independent and anti-rigid (i.e. their individuals can migrate to another phase type during their lifetime). *Role types* are anti-rigid (i.e. their individuals can migrate to another role type during their lifetime) and founded (i.e. depend on a particular pattern of associations). Therefore, individuals are modelled as existence dependent on natural types (cf. rigidity) and phase and role types are modelled as existence dependent of individuals (cf. anti-rigidity).

Using this conceptual backbone, various domain ontology concepts are categorised as natural, phase and role types. Humans are categorised as natural types (i.e. a human individual is a human for all of its life). Also movie types were categorised as natural types. A human's stages of life are considered phases (e.g. being a teenager does not require a particular association and may change to being an adult over time). Furthermore, all concepts of the ReWiRe ontology are considered role types as they all depend on their association with the context of pervasive applications (e.g. USER is the role of a HUMAN INDIVIDUAL in the context of pervasive applications, DEVICE is the role of a piece of hardware in the context of pervasive applications). Consequently, the conceptual backbone enables us to discriminate between the parts of the environment model that can change (i.e. anti-rigid) and those that cannot (i.e. rigid). For example, an adult movie will never be suitable for kids, but a person can grow up. Following the extension of the backbone with domain ontology concepts, the domain ontologies are extended with concepts specific for the intended application, which build an application ontology

[12] for parental control. Consequently, the application specific concepts of playing regular media content and adult media content are modelled as extensions of the SERVICE concept in the ReWiRe domain ontology [9].

Next to the categorization of domain and application ontology concepts, also the interactions between these concepts in our intended application need to be addressed. These interactions, which we consider as aspects [13] of ontology integration that crosscut multiple ontologies, are modeled as object lifecycle models. Fig. 2 shows one aspect of the interaction between the movie ontology and the media application ontology. It shows that 'all viewer' movies can only be accessed by the 'play media' service. A similar model has been created, representing that 'adult viewer' movies can only be accessed by the 'play adult media' service. The CR_INDIVIDUAL and END_INDIVIDUAL events in the finite state machine indicate that individuals (i.e. movies) can be assigned to each of these natural types.

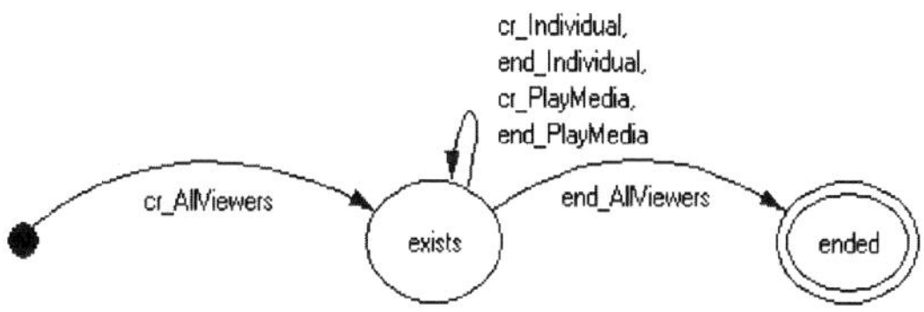

Fig. 2 All Viewers Movie x Play Media Service Aspect

Finally, we discuss the policy model for HUMAN INDIVIDUALS in the context of the parental control application. Fig. 3 shows that only adults (i.e. INDIVIDUALS of natural type HUMAN to which the ADULT phase type has been assigned) can have access to the PLAYADULTMEDIA service (i.e. CR_PLAYADULTMEDIA, end_PLAYADULTMEDIA) where other users only have access to the PLAYMEDIA service (i.e. CR_PLAYMEDIA, end_PLAYMEDIA).

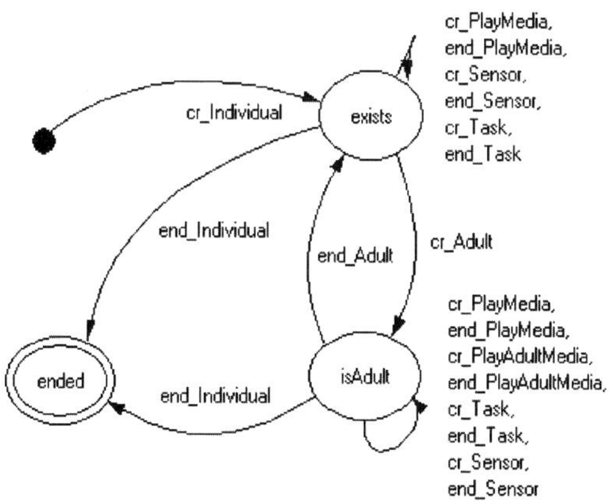

Fig. 3 Adults Only Policy

It should be noted that all finite state machines presented above were simplified for the reason of clarity. However, when these models are represented in the MERODE modeling tool (i.e. Mermaid[4]), the consistency checker automatically identifies e.g. orthogonal events, which can then be added to create the fully-fledged models.

4 Adopting MERODE Models in ReWiRe

The structural models produced by the MERODE design tool are transformed into a media domain ontology which is aggregated with ReWiRe's environment ontology as shown in fig. 4. An instance of this aggregated ontology describes the current environment context (i.e. available resources and relations that apply between them.

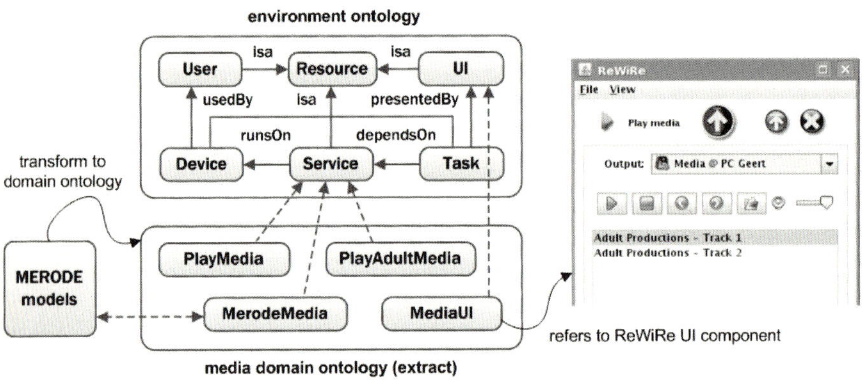

Fig. 4 Aggregated ontologies and media user interface component

We developed two services and a user interface which, together with the models, give rise to the pervasive media application. The PLAYMEDIA and PLAYADULTMEDIA services implement a media player component that exports a software interface to control a player from external services. The media player component publishes the player's state in the environment model and informs interested parties of state changes through sensor events (using ReWiRe's built-in notification mechanism). The PLAYMEDIA and PLAYADULTMEDIA services also shares media content residing on the device it runs on, by advertising it as MEDIACONTENT in the environment model. An additional service, the MERODEMEDIA service, fulfils a coordination role and implements a dedicated 'rewiring strategy' for the pervasive media scenario. Such a rewiring strategy aims to keep an application consistent at all times, in particular when the configuration of the environment changes. The coordination service is built on top of our previously designed

4 http://merode.econ.kuleuven.ac.be/mermaid.aspx

MERODE models. As such, the objects generated from the structural and behavioural knowledge base serve as a special-purpose model which is synchronized with the ReWiRe environment model using asynchronous events. For instance, the coordination service subscribes to R+ events (a new resource enters the environment), R- events (a resource leaves the environment) and player state events fired by a PLAYMEDIA or PLAYADULTMEDIA service and propagates this information to the MERODE models. Modifications to these models (e.g. transformations from one state into another one), either triggered by system or user events, are then translated into method calls that orchestrate a play media service.

Furthermore, we developed and deployed a migratable media user interface component as shown in figure 4. This interface presents available media content on end-user devices and features the option to select an output device to stream selected media to. In its back-end, this user interface leverages the MERODEMEDIA coordination service and its embedded behaviour models. For example, if a user selects adult content, MERODE models are traversed and if an adult check does not pass, the play button remains disabled in the user interface for the selected media.

5 Conclusions, Limitations and Future Research

This paper presented a framework, which uses ontology models that are connected to a conceptual backbone to provide pervasive applications with a shared and dynamic environment model, of which the (run-time) consistency is guaranteed by using the MERODE methodology. The presented parental control application is one of many potential scenarios that requires both real-world (e.g. the age of a person, the rating of a movie) and technological awareness (e.g. listing potential output devices). The integration of real-world and technological awareness in a changing (real-world and technological) environment is enabled by combining MERODE and ReWiRe.

Models that can capture the current state of a pervasive environment and its applications are vital components to build context-aware applications [14]. Ontologies have already been used to enable the development of pervasive applications [10, 15, 16]. For example, Preuveneers et al. proposed an ontology to capture the context of use of ambient intelligent environments in [17]. Chen et al. [15] designed a rich ontology for ubiquitous and pervasive applications (SOUPA) that is exploited in a broker-centric agent architecture to support knowledge sharing and context reasoning. In [10] context discovery and knowledge sharing are supported using an ontology-based context model and an OSGi-based middleware infrastructure. However, their ontology is mainly geared towards the creation and deployment of context-aware systems, and does not focus on runtime and thus (re)configuration support. As we do not only use ontologies to capture and query the execution context, but also to create dynamic pervasive applications which can be (re)configured at runtime. As the MERODE algebra has been used in model-driven design, it should be noted that the framework presented in this paper differs from modelling languages for pervasive systems (e.g. PervML[5]) by discriminating a dynamic environment model, of which the consistency is supported by the MERODE algebra, and a stable application design in ReWiRe, which interprets the environment model, where model-driven design approaches focus on the application design.

Since the use of the MERODE algebra has been limited to model-driven design, a tool that integrates the MERODE methodology and the ReWiRe framework is not available yet. Consequently, part of the integration between the MERODE and the ReWiRe environment had to be performed manually. In the future, the MERODE algebra will be integrated in the ReWiRe framework to automate environment model reconfiguration fully.

References

1. Geihs, K., Barone, P., Eliassen, F., Floch, J., Fricke, R., Gjorven, E., Hallsteinsen, S., Horn, G., Khan, M.U., Mamelli, A., Papadopoulos, G.A., Paspallis, N., Reichle, R., Stav, E.: A comprehensive solution for application-level adaptation. Softw. Pract. Exper. 39, 385–422 (2009)
2. Lobato, C., Garcia, A., Romanovsky, A., Lucena, C.: An aspect-oriented software architecture for code mobility. Softw. Pract. Exper. 38, 1365–1392 (2008)
3. Snoeck, M., Lemahieu, W., Goethals, F., Dedene, G., Vandenbulcke, J.: Events as atomic contracts for component integration. Data & Knowledge Engineering 51, 81–107 (2004)
4. Snoeck, M., Dedene, G.: Existence dependency: The key to semantic integrity between structural and behavioral aspects of object types. IEEE Transactions on Software Engineering 24, 233–251 (1998)
5. Dedene, G., Snoeck, M.: Formal deadlock elimination in an object oriented conceptual schema. Data & Knowledge Engineering 15, 1–30 (1995)
6. Snoeck, M., Dedene, G.: Generalization/specialization and role in object oriented conceptual modeling. Data & Knowledge Engineering 19, 171–195 (1996)
7. Snoeck, M., Poels, G.: Improving the Reuse Possibilities of the Behavioral Aspects of Object-Oriented Domain Models. In: Laender, A.H.F., Liddle, S.W., Storey, V.C. (eds.) ER 2000. LNCS, vol. 1920, pp. 423–439. Springer, Heidelberg (2000)
8. Snoeck, M.: Object-oriented enterprise modelling with MERODE. Leuven University Press, Leuven (1999)
9. Vanderhulst, G., Luyten, K., Coninx, K.: ReWiRe: Creating interactive pervasive systems that cope with changing environments by rewiring. In: 4th International Conference on Inteligent Environments, pp. 1–8. IEEE, Seattle (2008)
10. Gu, T., Pung, H.K., Zhang, D.Q.: Toward an OSGi-Based Infrastructure for Context-Aware Applications. IEEE Pervasive Computing 3, 66–74 (2004)
11. Parsons, J., Li, X.: An ontological Metamodel of Classifiers and Its Application to Conceptual Modelling and Database Design. In: Parent, C., Schewe, K.-D., Storey, V.C., Thalheim, B. (eds.) ER 2007. LNCS, vol. 4801, pp. 214–228. Springer, Heidelberg (2007)

---

[5] http://oomethod.dsic.upv.es/labs/index.php?option=com_content&task=view&id=40&Itemid=77

12. Guarino, N.: Formal Ontology and Information Systems. In: Proceedings of FOIS 1998, pp. 3–15. IOS Press, Trento (1998)
13. Kiczales, G., Lamping, J., Mendhekar, A., Maeda, C., Lopes, C., Loingtier, J.M., Irwin, J.: Aspect-oriented programming. In: Aksit, M., Matsuoka, S. (eds.), Jyvaskyla, Finland, pp. 220–242 (1997)
14. Dey, A.K.: Understanding and Using Context. Personal and Ubiquitous Computing 5, 4–7 (2001)
15. Chen, H., Perich, F., Finin, T., Joshi, A.: SOUPA: standard ontology for ubiquitous and pervasive applications. In: The First Annual International Conference on Mobile and Ubiquitous Systems: Networking and Services, 2004. MOBIQUITOUS 2004, pp. 258–267 (2004)
16. Peters, S., Shrobe, H.E.: Using semantic networks for knowledge representation in an intelligent environment. In: First IEEE International Conference on Pervasive Computing and Communications, pp. 323–329. IEEE, Los Alamitos (2003)
17. Preuveneers, D., Van den Bergh, J., Wagelaar, D., Georges, A., Rigole, P., Clerckx, T., Berbers, Y., Coninx, K., Jonckers, V., De Bosschere, K.: Towards an Extensible Context Ontology for Ambient Intelligence. Ambient Intelligence, 148–159 (2004)

Toward a Behavioral Decomposition for Context-Awareness and Continuity of Services

Nicolas Ferry, Stéphane Lavirotte, Jean-Yves Tigli,
Gaëtan Rey, and Michel Riveill

Abstract. Many adaptative context-aware middleware exist and most of them rely on so-called vertical architectures that offer a functional decomposition for context-awareness. This architecture has a weak point: it does not allow the system handling both dynamics of the changing environment and applications. To avoid this, we propose an approach for context-awareness based on a behavioral decomposition, and because each behavior must complete all functionalities necessary for context-awareness, we introduce an hybrid decomposition. It consists in a functional decomposition into a behavioral decomposition. This approach derives benefits from both decomposition, first allowing to handle environment and application's dynamics, second introducing reusability and modularity into behaviors.

1 Introduction

Nowadays, with the miniaturization of computer hardware, many objects with computational capabilities are dissolving in our daily lives. Thus, the idea of personal computer as the single smart object or as a universal digital assistant is fading away.

Ambient systems consist of two categories of entities: (1) living entities, as the user, and (2) systems with computational capabilities. A system relies on a hardware

Nicolas Ferry
I3S (UNS - CNRS) and CSTB, 290 route des Lucioles - BP209 06904 Sophia-Antipolis France
e-mail: nicolas.ferry@unice.fr

Stéphane Lavirotte
I3S (UNS - CNRS), 930 Route des Colles - BP 145 06901 Sophia-Antipolis France
e-mail: stephane.lavirotte@unice.fr

Jean-Yves Tigli · Gaëtan Rey · Michel Riveill
I3S (UNS - CNRS)
e-mail: tigli@polytech.unice.fr, gaetan.rey@unice.fr, riveill@unice.fr

J.C. Augusto et al. (Eds.): ISAmI 2010, AISC 72, pp. 55–62.

infrastructure which can provide a software infrastructure. Under these constraints, the editable software part of the system allows to implement new functionalities for new applications.

Ambient systems are characterized by using devices and objects of everyday life, and can take many forms. The technological heterogeneity of those devices and objects is still the biggest challenge to overcome in order to enable them to interact with the system. Moreover, we must consider the semantic heterogeneity of these entities that can be introduced into ambient system. Another major feature of ambient system is the high variability of software infrastructure. These infrastructures evolve dynamically led by appearances and disappearances of objects or devices. The topologies of this infrastructure are dynamic due to arbitrary node mobility, failures or energy saving.

Because environment's nature is highly variable, even the corresponding software infrastructure, pervasive systems have to handle those variations and offer such dynamicity. These systems have to be able to adapt to their context; they must be context-aware. And adaptations to these changes have to be done quickly to follow those variations but not to often to preserve the application's stability and not to divert the user.

2 Related Works: Context and Context-Awareness

Generally and specially in the field of ambient computing, the context appears as the supplement of all editable software entities. So this supplement can be:

- All hardware or software infrastructures,
- The environment,
- All the entities without any computational capabilities like users or raw objects.

2.1 Context-Awareness

Classical mechanisms for context-awareness rely on a functional decomposition to provide reusability and evolution facilities. This decomposition is usually based on key functionalities [6]. The first stage of sensing is to gather contextual information also called observables. These observables are then transformed as symbolic observables about the state of the environment, for example using ontologies.

Middleware such as the Context Toolkit [9], SOCAM [11], Contextors [14], providing mechanism to collect, store and sometimes process these observables, have emerged. They offer some mechanisms to discover heterogeneous entities, but also for deduction and filtering of data collected using some centralized mechanisms. However, some middleware as COWSAmI [1] or Construct [7] are based on a decentralized architecture.

Context observation is a sub-problem of context-awareness. Indeed, as shown in SAFRAN [8], application's behavior adaption is also a crosscutting concern and a part of context-awareness mechanism [3]. Hence, as a result of contextual

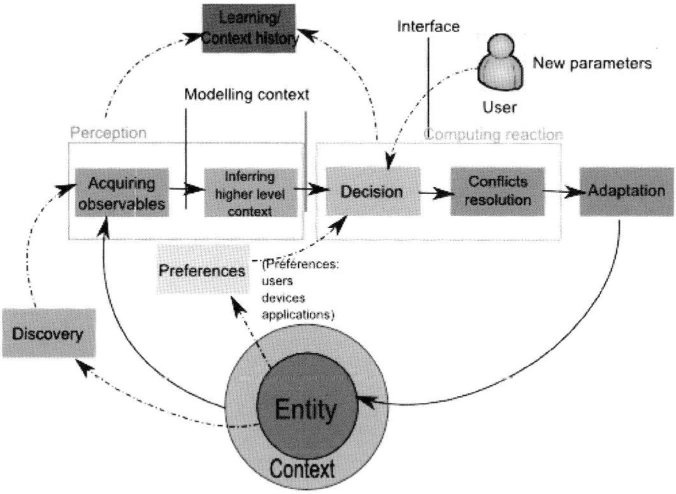

Fig. 1 Functional decomposition of context-aware mechanism

information collection, data will be used during a decision stage, also called situation identification stage. It will produce an action plan which will be implemented by the adaptation mechanism also called the control mechanism. Of course, these steps can be refined as shown in Fig 1.

2.2 Limitations

The functional decomposition often results in the development of so called "vertical" architecture, consisting in a set of various layered functionalities.

Most of context-aware middleware are based on those architectures. They have a weak point: the data centralization in at least one of the functionality of the decomposition. This means that they rely on a common representation or a global model of context. For example, in Gaia [15], a context-aware middleware, observables are stored in a single entity: the "context file system". We found the same problem with SOCAM [11] or CARMEN [2]. For example, CARMEN uses a specialized LDAP containing user profiles for the decision stage. However, centralizing data remains hardly imaginable in an ambient system where you have to manage a throng of heterogeneous entities and data. In fact the environment is not known a priori and is constantly evolving, so considering the whole environment is similar to considering the universe, which is not possible. A centralized approach involving potentially a throng of information then comes with a bottleneck of performance and does not allow the system to scale.

But these architectures for context-awareness have a central role in an ambient system since they interact with both environment and applications. Then, they should respect application and environment's dynamics. So reactivity is a key concern of ambient computing, both for adaptation triggering and for adaptation time.

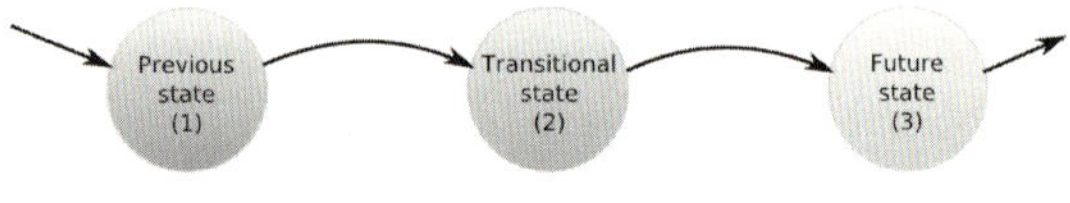

Fig. 2 The three states of an adaptive application

But reactivity defined as follows can hardly be proposed by middleware using a vertical architecture.

We consider that adaptive applications are always in one of the three states presented in Fig 2. States (1) and (3) are normal execution states of the application, where it is consistent with its environment. It means that the application's behavior is based on what is relevant in its environment and this is the expected behavior for a particular situation. During the transitional state (2) the application is in its adaptation phase and unavailable. It is considered in an inconsistent state because the application is not in line with its environment.

Consequently, Adaptation's dynamic has to be consistent with the dynamic of the changing environment [10]. In other words, it is essential that:

- The system is not unavailable (2) for too long while adapting, so adaptation has to be as fast as possible in order to obtain a consistent application (3). Otherwise, the system could become unstable and may never reach a normal execution state before new evolutions occur in its environment.
- The system does not stay in the previous state (1) too much time before reacting to environment changes.

Moreover, adaptation's dynamic has to be consistent with application and/or user operation (application's dynamic). In other words, it is essential that:

- The system does not go too frequently from state (1) to (2) and produce an unstable and inconsistent application.
- Too many lag can disrupt and divert the user from the system [12].

Vertical architectures does not allow handling all the dynamics described above. Indeed system reactions to change in context depend on the reactivity of each part of treatment (each layer). Changes are all evaluated simultaneously using the same processed data. This leads to a system with a fixed dynamic.

Therefore, it appears that the complexity of processing and representation of contextual information impact the dynamic of the mechanism of context-awareness. Then it is necessary to study another kind of decomposition to handle these dynamics. This line of study is not unlike the general approach to problem solving by a human operator. Indeed the Rasmussen's model [13] called SSRK defines three levels of cognitive processes:

- Skill Based: the skill level corresponds to actions that are done automatically.
- Rule Based: the rules' level corresponds to procedures (algorithmic) implemented by human to define its action.
- Knowledge Based: the level of declarative knowledge implements a process that requires many cognitive resources and so that take time to make a decision and implement its action.

We then see that it is a decomposition of various processes with their own dynamic working separately.

3 Reactivity and Behavioral Decomposition of Context-Awareness Mechanism for Application's Adaption

Similar architectures exist in the field of robotics, where they are called "horizontal" architecture. The aim of this approach is to specialize a generic minimal core of an application with some specific modules called horizontal layers [16]. These layers are independent of each other and work in parallel. Each layer is connected to the world via a set of sensors and can act through actuators. This kind of architecture introduces a new type of decomposition: the *behavioral decomposition*.

According to Bryson [5], the behavioral decomposition is an architectural approach that decomposes the intelligence in terms of behavior such as eating or walking, rather than generic processes like planning or observing. A behavior corresponds to an horizontal layer and activities may consist of sets of managed behavior. A major contribution of these approaches is not to see the system as a sequence of processes and thus functionality, but rather as a parallelization of processes which can produce a coherent activity. More complex behaviors are achieved with simple ones like for divide and conquer strategies. These architectures comply to the following characteristics:

- Each layer captures relevant information from the environment, sensing the surrounding to a degree sufficient to achieve the necessary [4].
- There is no need for a representation of the environment in a decentralized approach: "The world is its own best model" [4]. The only accurate environmental data are obtained immediately by the sensors.
- They are based on many small behaviors of low complexity for the best possible reactivity.

Horizontal architectures need a coordination mechanism to combine the output data of each layer in order to obtain a rational and coherent global behavior. A simpler mechanism induces more reactivity.

In the field of context-awareness, this behavioral decomposition relies on autonomous and separated processes of context-awareness. They produce adequate changes in the application from relevant gathered information in order to provide a coherent global behavior to the application. We call such processes a *basic adaption to context behavior* (BACB).

This kind of architectures enables a better control of environment and application's dynamics. Indeed, treatment directed by a BACB is made only when necessary. Moreover, it is as fast as possible since each BACB computes only what is relevant to them. In fact, because these BACB are independent of each other, they have their own dynamic which is not anymore dependent of other treatments. Then we are able to write BACB with various levels of complexity to best fit the dynamics imposed by the environment and the application. Indeed, we will be able

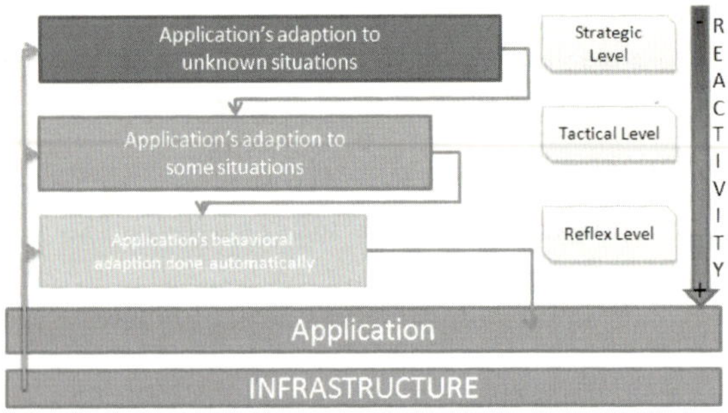

Fig. 3 The proposed CONTINUUM architecture. This architecture relies on three levels. Levels closest to the infrastructure are the most reactive. These levels are similar to the three levels of SSRK. The level N is based on the mechanisms offered by the level N-1 and may act on it.

to study for each BACB their dynamics (time of response) and their compatibility with the evolving environment (not enough or too reactive). On the other hand, we can also study interactions between these behaviors and their good management. Each BACB is scheduled with a late manager, it may be a mechanism to resolve adaptation conflicts.

In the CONTINUUM French national project, we propose a context-aware mechanism for continuity of services in ambient systems, see Fig 3. This architecture is based on a behavioral decomposition. And each level has its own dynamic according to their priority and the complexity of their process.

4 Towards an Hybrid Approach

As we saw earlier, a BACB gather some contextual information and act on it in return. A behavior identifies an action to perform according to a relevant context, so it must respect the classical steps of the functional decomposition: perception, decision, and reaction. Such a BACB can then be decomposed into several functionalities. So we introduce an hybrid approach consisting of a functional decomposition into a behavioral decomposition. A behavior is itself composed of a set of functionalities that, applied to context-awareness, spreads from perception to reaction.

The hybrid approach derives some benefits from both decomposition, first to control its dynamics and independence thanks to behavioral decomposition, secondly improving reusability and modularity into behaviors thanks to functional decomposition. More precisely, hybrid decomposition offers, compared to behavioral decomposition, modularity into behavior and therefore maintenance facilities and thus evolving facilities. Moreover it induces a better separation of concerns into behaviors. The modularity introduced by functional decomposition could enable adaption of behaviors. For example, in the CONTINUUM project, a scenario applied to the hydrant man job defines a set of services to help them. One part of the job is to stop water valves when necessary. In this project a service of the system indicates, in the car, the localization of those valves. According to the brightness, these information are visual or voiced. Moreover, the system help them to manage their intervention (description, ...) from their car. This is not allowed when the car moves too fast.

In this small part of the scenario, we can define three behaviors: (1) visual or (2) voiced indications of the valve's localization, (3) intervention management. We can see that those three behaviors must respect various dynamics. So, blocking the use of the help system is critical, it has to be done quickly (environment's dynamic). Otherwise changing the kind of interaction with the localization service must be done more slowly not to pass from an interaction to another incessantly (application's dynamic). Here appears the advantages of the behavioral decomposition. On the other hand, the perception module of behavior (1) and (2) can be reused, thanks to the functional decomposition included in the hybrid decomposition.

Then it appears that this approach requires an inversion in design methodology of mechanisms provided by a middleware. Indeed, it is not to decompose behaviors in a set of features but to write a set of behavior from reusable features. Thus, the major stages of designing an hybrid architecture are: (1) identifying behaviors, (2) specifying their inputs and outputs, (3) identifying features, (4) specifying the feature chaining, (5) implementation. When creating behavioral architectures, specification of inputs / outputs is particularly important. Unlike in classical functional decomposition, aggregation of data is not given as a pre-requisite of the system (hence the need for data representation in functional approaches) but between behavior's outputs.

5 Conclusion

Many context-aware middleware exist and most of them rely on vertical architectures that offer a functional decomposition for context-awareness. This architecture has a weak point: it does not allow the system to handle both dynamics of the changing environment and applications. Handling these dynamics is a key concerns in the field of ambient intelligence since ambient systems must consider a throng of informations and devices. To avoid this, we proposed in this paper to adapt an approach from robotics called behavioral decomposition to context-awareness. To do this, we introduced the concept of basic adaptation to context behavior (BACB). Because each BACB relies on the classical steps found in the functional decomposition from perception to reaction, we introduce an hybrid decomposition. It consists in a functional decomposition into a behavioral decomposition. This approach derives benefits from both decomposition, allowing to handle environment and application's dynamics, and introducing reusability and modularity into behaviors.

Acknowledgements. This work is part of the Continuum Project (French National Research Agency) *ANR-08-VERS-005*.

References

1. Athanasopoulos, D., Zarras, A., Issarny, V., Pitoura, E., Vassiliadis, P.: CoWSAMI: Interface-aware context gathering in ambient intelligence environments. Pervasive and Mobile Computing 4(3), 360–389 (2008)

2. Bellavista, P., Corradi, A., Montanari, R., Stefanelli, C.: Context-aware middleware for resource management in the wireless Internet. IEEE Transactions on Software Engineering 29(12), 1086–1099 (2003)
3. Bottaro, A., Bourcier, J., Escoffier, C., Lalanda, P.: Context-aware service composition in a home control gateway. In: International Conference on Pervasive Services, pp. 223–231 (2007), http://doi.ieeecomputersociety.org/10.1109/PERSER.2007.4283920
4. Brooks, R.: Elephants Don't Play Chess. Designing Autonomous Agents: Theory and Practice from Biology to Engineering and Back, 3–15 (1991)
5. Bryson, J.: Intelligence by design: Principles of modularity and coordination for engineering complex adaptive agents. Ph.D. thesis (2001)
6. Coutaz, J., Crowley, J., Dobson, S., Garlan, D.: Context is key. Communications of the ACM 48, 49–53 (2005)
7. Coyle, L., Neely, S., Stevenson, G., Sullivan, M., Dobson, S., Nixon, P., Rey, G.: Sensor fusion-based middleware for smart homes. International Journal of Assistive Robotics and Mechatronics 8(2), 53–60 (2007)
8. David, P., Ledoux, T.: Towards a framework for self-adaptive component-based applications. LNCS, pp. 1–14 (2003)
9. Dey, A., Salber, D., Futakawa, M., Abowd, G.: An architecture to support context-aware applications. Submitted to UIST (1999)
10. Ferry, N., Hourdin, V., Lavirotte, S., Rey, G., Tigli, J.Y., Riveill, M.: Models at Runtime: Service for Device Composition and Adaptation. In: MRT 2009, p. 10 (2009)
11. Gu, T., Pung, H., Zhang, D.: A service-oriented middleware for building context-aware services. Journal of Network and Computer Applications 28, 1–18 (2005)
12. MacKenzie, I., Ware, C.: Lag as a determinant of human performance in interactive systems. In: Proceedings of the INTERACT 1993 and CHI 1993 conference on Human factors in computing systems, pp. 488–493. ACM, New York (1993)
13. Rasmussen, J.: Information processing and human-machine interaction: An approach to cognitive engineering. Elsevier Science Inc., NY (1986)
14. Rey, G., Coutaz, J.: The Contextor Infrastructure for Context-Aware Computing. In: ECOOP 2004, Workshop on Component-Oriented Approach to Context-Aware Systems, Citeseer (2004)
15. Román, M., Hess, C., Cerqueira, R., Ranganathan, A., Campbell, R., Nahrstedt, K.: Gaia: a middleware platform for active spaces. ACM SIGMOBILE Mobile Computing and Communications Review (2002)
16. Zhang, C., Jacobsen, H.: Resolving feature convolution in middleware systems. ACM SIGPLAN Notices (2004)

A User-Oriented and Context-Aware Service Orchestration Framework for Dynamic Home Automation Systems

Thomas Coopman, Wouter Theetaert, Davy Preuveneers, and Yolande Berbers

Abstract. The home environment has become increasingly sophisticated. Sensors and actuators allow a house to be *programmed* so that it can intelligently adapt to user needs. Unfortunately it has not become any easier for the user to manage all these devices. Our ontology driven framework describes a user-oriented automation system to help with this process by not only orchestrating home automation devices, but also by integrating new equipment and automating them as much as possible. By combining devices with a common functionality in a given context, relevant devices are pulled together in an intuitive user interface in which one can express preferences at a high level of abstraction. The big advantage is that our framework allows for the home environment to be adjusted without the user having to know about the technical details of devices in the home automation system.

Keywords: home automation, context-awareness, service composition.

1 Introduction

Computers are becoming smaller and more powerful. The image of the computer only used as a desktop is becoming obsolete [12]. Today computers are everywhere, even in our home environment: lighting controlled by light sensors, garage doors that open and close automatically, curtains that close automatically in the evening, time triggered coffee makers. We see that technological advances have equipped household appliances with more and more functionality and flexibility. The other side of the medal is that the manuals on how to use home automation systems are

Thomas Coopman · Wouter Theetaert
Department of Computer Science, Celestijnenlaan 200A, B-3001 Heverlee, Belgium
e-mail: `firstname.lastname@student.kuleuven.be`

Davy Preuveneers · Yolande Berbers
Department of Computer Science, Celestijnenlaan 200A, B-3001 Heverlee, Belgium
e-mail: `firstname.lastname@cs.kuleuven.be`

J.C. Augusto et al. (Eds.): ISAmI 2010, AISC 72, pp. 63–70.

becoming bigger and bigger. It is a daunting task to understand all the possibilities and capabilities. Furthermore, many sensors or actuators will need to work together, and configuring such installations will not be easy. Collecting and managing context information [6] from the vicinity is becoming more and more practical every day. If context information were also to be used during the interaction between a person and the home automation system, then the human-machine interaction would become a more user friendly process [10].

In this paper we describe an ontology-driven system that addresses the above concerns. Our dynamic, context-sensitive composition framework leverages service orchestration [8] to improve the interoperability between various devices in a home automation system and provides an interface that will make life easier for the end user. Sensors and actuators are enhanced with a semantic description of their capabilities. The framework uses context information to offer management and automation facilities to the user at the highest level of abstraction possible. In this way the user has an intuitive interface to adjust the system to his own preferences.

After discussing related work in section 2, we continue in section 3 with a motivating scenario illustrating how our context-aware service orchestration framework for home automation works. In section 4 we discuss the semantic representation and mapping in our framework. The architecture of the framework is described in section 5. We end with some conclusions and topics for future work in section 6.

2 Related Work

Before we dive into the design of our framework, we first discuss existing systems that use context-sensitivity and/or service composition as key building blocks. We highlight potential concerns and discuss elements reused in our own framework.

In [2], Chen and Kotz present an overview of applications that use context-sensitivity to their advantage. There are applications that forward phone calls to a device the closest to the user, tourist guides that offer a word of explanation about artefacts in the vicinity [4], and several others. Many applications, however, do not make sufficient use of the offered possibilities. They confine themselves mainly to the use of information about location and time. First, simple applications respond to context sensors that capture the presence of people in the environment, temperature, light intensity, sound, etc. Our system needs to infer complex situations from the low level context (e.g. *a person is in a meeting room, there are other people present, there is noise* could mean there is a meeting in progress). Second, context-sensitive systems often use a traditional application model, where the context dependencies are modeled with a lot of *if-then-else* rules that specify how the program must respond to a particular input.

The above approach has obvious disadvantages: when new context sensors or actuators are added in the home environment, old rules may have to be rewritten as some of the rules may now conflict with one another or have become ambiguous. Also, the approach is not flexible for a home automation system as the end user is not a developer and he should not be one. The user needs to set his goals in a

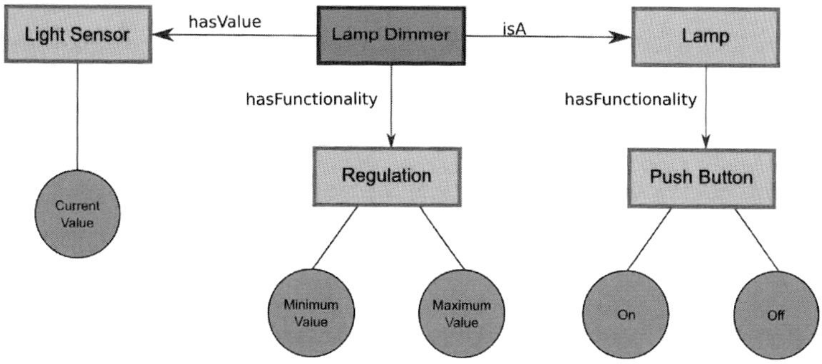

Fig. 1 A graphical representation of an ontology for a lamp dimmer

simple way, without technical knowledge of the various services he uses. As such, user intentions should be declarative: the user should state what should be achieved (e.g. moderate lightning), but not how this should happen (e.g. change the light intensities of the individual light bulbs). With each component being semantically enhanced, it is also much easier for a new device to be automatically integrated into the system. The home automation system presented here uses many context elements with ontology- and rule-based inference to dynamically adjust the composition of sensors and actuators, by automatic translation of highlevel goals into low-level device orchestrations and vice versa. In the literature different ontologies to describe context and services have been proposed. For representing context, we have used and extended SOUPA [3]. Two ontologies that are useful to model services within the home automation environment are Domo-ML [11] and DogOnt [1]. These ontologies are quite complete, but lack some useful information to enable dynamic composition. In traditional service composition systems, such as in web services [7, 9], the interconnection of two blocks is often based on a simple input / output agreements. To make sure that only compatible devices are linked, their descriptions need to match semantically. The DogOnt ontology has been extended for this purpose. Fig. 1 shows how a lamp dimmer in DogOnt can look like.

3 Motivating Scenario and Requirements

In this section we describe how our framework operates by means of a motivating scenario. Suppose Alice and Bob recently bought a home automation system that uses our framework. Upon startup, the system will be searching for devices (sensors, actuators and controllers) present in the house. When Bob arrives home from work, he likes to watch the news headlines on television. He takes the remote control with on display and connects to the automation system. Since this is a new system, there are no compositions set. Bob chooses to create a new composition, and will see a list of potential building blocks for the composition. Bob chooses the categories *light*, *video* and *video* he wants to configure when he wants to watch

Table 1 Services with a similar functionality

Service	Category	Service	Category
Spot lights	Light	Shutters	Security
Reading lamp	Light	Door	Security
Light sensor	Light	Radio	Music
Shutters	Light	CD player	Music

the news (see Fig. 2). Bob wants moderate light, neutral sound and the news on his favorite channel. The system will convert the high-level settings of the user to specific low-level instructions for the various devices. Moreover the system ensures that the devices are linked together. If Bob would move, the composition follows him along. So the video of the news is moved to the nearest screen, and the sound to the closest sound system in the neighborhood. Alice comes home after having bought a new reading lamp for the living room. The system detects the light and – based on the lamp's semantic description – adds it directly to the category *light*. When Bob wants to watch the news afterwards, he can recall the composition that he just made. The light setting *moderate* will be re-activated and the settings of the lamps will be adjusted for the fact that a new lamp is added.

All resources should be classified according to their semantic description into blocks with similar functionality (see Table 1). Thus, all lighting devices should be classified in the right category *Light*. Note that certain services can belong to different categories: e.g. the shutters can serve a purpose when controlling the lightning, but also for the security of the house. The list of building blocks shown to the user should be context-dependent: the block *audio* should only be displayed when Bob is in a room in which devices are present that can produce sound. Each of the selected blocks can be configured separately at a high level. The system described above should include some special features. It should make use of information from the environment: the system must show only blocks that have some utility in the current context. The user must also have no knowledge of the various individual devices in the system. He should be able to enter his preferences at a high level and the system should make the desired composition. There may of course be some predefined compositions in the system, but the user should have sufficient possibilities to customize everything to his desires.

4 Mapping of User Settings to Low-Level Instructions

It is now clear that our framework needs to automate a lot of things without user intervention. The intention is that a user can give high level goals while the framework converts them into low-level instructions. To meet these requirements we need to represent context and services to characterize the various possibilities of the home automation system. Such representation must meet two conditions:

- First, the representation must be sufficiently expressive and extensible: different services must be linked and this in a 'smart' way. Furthermore, services may

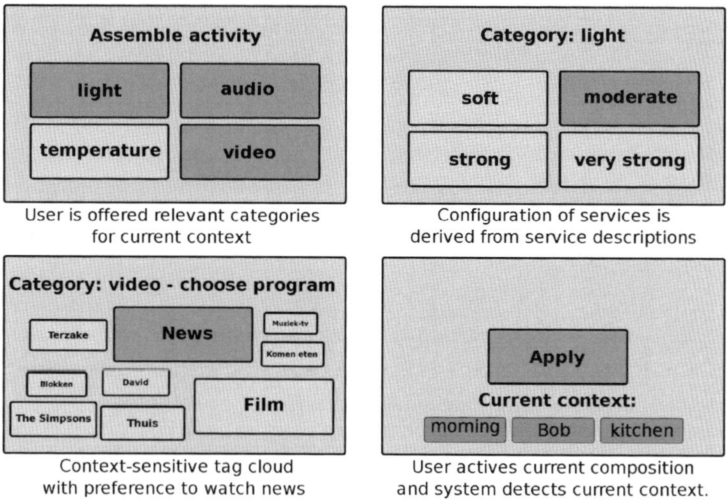

Fig. 2 Illustrations of the interface of the framework to the end user

come and go as old devices are being replaced. This is more than merely linking two blocks where the input of one block corresponding to the output of another block. The link must is also be semantically correct. It makes no sense to control the lightning with a temperature sensor.

- The second condition is that the representation must allow for information reasoning to model common sense knowledge about home automation systems. Such a system that works with sensors must assume that people may sometimes provide incomplete or incorrect information. Conflicts that may arise have to be resolved by the context model.

Due to these requirements, we have extended our context ontology [5] and also use an ontology based model (see Fig. 1) to model home automation knowledge and information. Our framework relies heavily on subsumption relationships between classes and properties in ontologies to detect incompatibilities. Common sense knowledge is expressed with semantic rules that are processed by the HermiT Web Ontology Language (OWL) reasoner[1]. For example, during discovery discovery we may be looking for a service that will need to interact with another given service. Therefore, the interfaces of the cooperating services should match. This is illustrated in the service composition in Fig. 3. Service 2 delivers its output of type B to service 3 and requires input of type A from service 1. Any instance i_1 (here of type X) whose type is subsumed by type A will match. Likewise should the type of instance i_2 (here of type B) be subsumed by type Y. We use subsumption in a similar way to find devices with specific capabilities.

[1] http://hermit-reasoner.com/

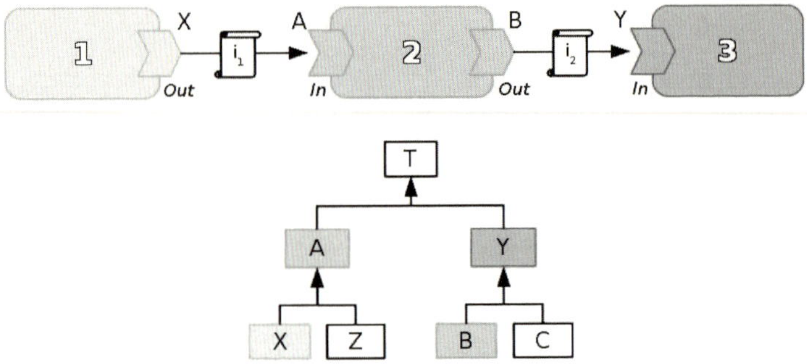

Fig. 3 Semantically matching input/output interfaces and categories in a service orchestration

An important aspect of the implementation is the conversion between high- and low-level contextual values. First, the low-level configuration possibilities of the devices should be presented to the end-user in a easy way (see Fig. 2). For that reason, the concept 'mapping' was introduced. In the ontology, each block is connected with a specific mapping. This mapping determines how the end-user can set the values for that particular block. Second, the high-level configurations of the end-user have to be enforced at the level of the devices. When an end-user asks for 'moderate light', this should be translated into practical values of the various lighting devices in the room. While doing this, the actual value of a certain property should be taken into account: when 'moderate light' is asked and this matches with some lighting value, the actual light value in the room should be measured before deciding whether the lights or the shutters should be activated.

5 Architectural Overview of the Home Automation Framework

The architecture of the system consists of several components as can be seen in Fig. 4. On the figure are two major parts: the internal system on the left, and external entities (services, context sensors, users via user interfaces) on the right. All communication passes through the *Communications Manager*. It communicates with all sensors and actuators in their own protocol and ensures that the other devices should not care about different protocols being in use. There are three types of information flows that enter the system: a device that sends information about its services (e.g. coffee), context sensors that provide information about the current situation (e.g. a light sensor), or users that provide information through a user interface (e.g. an interactive display in the house) to change or activate some settings.

The information from the sensors (the low-level context) is sent to the *Context Manager* which reasons and infers high-level context, which is then forwarded to the *Composition Manager*. For example. a Bluetooth sensor reports the presence

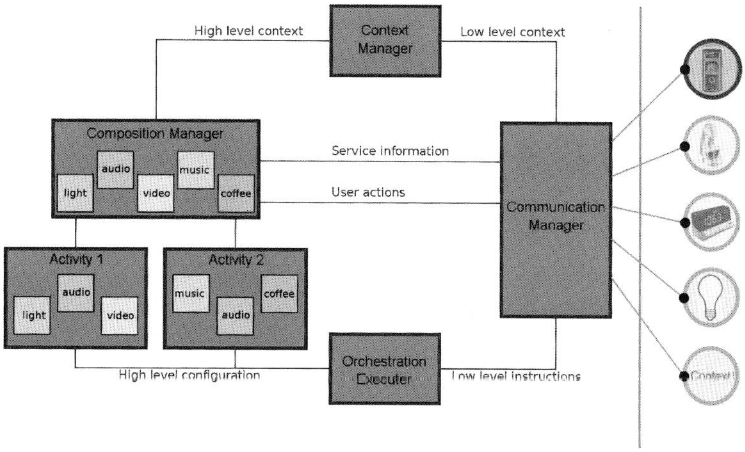

Fig. 4 Architectural overview of the home automation system

of a smartphone. The *Context Manager* then finds where the sensor is located (the living room) and who owns the smartphone (Bob). This way high-level context information is derived (Bob is in the living room).

Besides high-level context information, the *Composition Manager* also obtains information about the various services in the system, and this in the form of ontologies. As mentioned earlier, the services are grouped into blocks with similar functionality (e.g. light, audio, video, etc.). The user can use these blocks to assemble a new composition. The *Composition Manager* helps by just showing the blocks that are currently relevant (those whose devices are currently available).

Finally, the composition needs to be enforced for the various services. The *Orchestration Executer* is responsible for converting the information of the various blocks into the service specific implementations. This conversion is only possible if the ontologies of the services contain enough information about their control and data flow.

6 Conclusions and Future Work

This paper addresses the concern of managing complex dynamic home automation systems. More and more devices enter the home environment, and it is becoming increasingly more difficult to install and use them. Moreover is not evident for several sensors and actuators to work together out of the box. Furthermore, not every connection makes sense (you don't want to control the heating with a light sensor). In addition, we want to enable the user to manage the composition rather than the manufacturer of the system. Therefore, it is obvious that managing the home automation system should be done in an intuitive way. We described a motivating scenario and examined to what extent existing solutions could help. Many existing context-sensitive systems do not make sufficient use of context and are implemented in a very static way. In almost all service composition tools, the orchestration is carried out by a developer based on simple input / output match. This is inconsistent with our expectations that the composition should be configured declaratively by the end user rather than in an imperative manner by the developer. Finally, we discussed the architecture of our context-aware service orchestration framework that supports such an approach for dynamic home automation systems.

As future work, we envision extensions to the system in the form of some predefined composition templates that can easily be changed by the user. These could be useful for modeling recurring circumstances, for example if nobody is in the house (the lights are off, all devices are switched off, the alarm is set). Furthermore, we are also thinking of extending our framework with a learning component so that the compositions are not only explicitly defined by the end users, but can be implicitly defined according to learned patterns in user behavior.

References

1. Bonino, D., Corno, F.: Dogont - ontology modeling for intelligent domotic environments. LNCS, pp. 790–803 (2008)
2. Chen, G., Kotz, D.: A Survey of Context-Aware Mobile Computing Research, Department of Computer Science. Dartmouth College (2000)
3. Chen, H., Finin, T., Joshi, A.: The soupa ontology for pervasive computing. Ontologies for Agents: Theory and Experiences, 233–258 (2005)
4. Cheverst, K., Mitchell, K., Davies, N.: Design of an object model for a context sensitive tourist guide. Computers & Graphics 23, 883–891 (1999)
5. Preuveneers, D., et al.: Towards an extensible context ontology for Ambient Intelligence. In: Markopoulos, P., Eggen, B., Aarts, E., Crowley, J.L. (eds.) EUSAI 2004. LNCS, vol. 3295, pp. 148–159. Springer, Heidelberg (2004)
6. Dey, A.K.: Understanding and using context. Personal Ubiquitous Comput. 5(1), 4–7 (2001)
7. OASIS. Web services business process execution language version 2.0 (2007), http://docs.oasis-open.org/wsbpel/2.0/wsbpel-v2.0.html
8. Papazoglou, M.P., Georgakopoulos, D.: Service oriented computing. Commun. ACM 46(10), 24–28 (2003)
9. Rao, J., Su, X.: A survey of automated web service composition methods. In: Proceedings of the First International Workshop on Semantic Web Services and Web Process Composition, pp. 43–54 (2004)
10. Schmidt, A.: Implicit Human Computer Interaction Through Context. Personal and Ubiquitous Computing 4(2/3) (2000)
11. Sommaruga, L., Perri, A., Furfari, F.: Domoml-env: an ontology for human home interaction. In: SWAP 2005: Proc. of the 2nd Italian Semantic Web Workshop, p. 166 (2005)
12. Weiser, M.: The computer for the 21st century. Scientific American 265(3), 94–104 (1991)

A Multi-user Smart System for Reminding Medication

Rui Pedro Marques

Abstract. This paper proposes architecture for a multi-user system for reminding medication, especially aimed at older people since they are those who have more difficulty managing their medication. Throughout the paper the proposed system and its key functionalities are described. Furthermore, the main research questions associated with the projected system are enumerated; and a research plan which will allow those issues to be addressed and which will evaluate the feasibility of the idea is also presented. Finally the methodology which is intended to be put into practice is indicated and how the research results can be obtained is also defined.

Keywords. ambient assisted living, managing medication, older people, pervasive computing, ubiquitous computing.

1 Introduction

People are living longer and most of older adults are hoping to remain in their homes as long as possible. Fortunately, many technologies have the potential to help older adults maintain their independence and health [1]. The independence of the elderly includes, beyond other aspects, taking the right medication in the right dose at the right time.

A reminder function is an important topic within managing medication subject, because besides patients have to take the correct dosages of medicines, it is relevant to take them at the right time and in certain correct intervals of time. It has been estimated that about 50% of the prescriptions filled are not correctly taken and 55% of the elderly fail to comply with their medication regimens, with 26% of those errors being potentially serious. The errors in taking medication (omissions or delays in the correct dose or taking wrong dosage) have impacts on the patient's

---

Rui Pedro Marques
Universidade de Aveiro
Campus Universitário de Santiago 3810-193 Aveiro, Portugal
e-mail: `ruimarques@ua.pt`

J.C. Augusto et al. (Eds.): ISAmI 2010, AISC 72, pp. 71–78.

health but also negative economic effects because they are responsible for 10% of hospital admissions and for 23% of nursing home admissions [2, 3].

Several challenges are associated with the resolution of such problems, namely: providing mechanisms to reduce the significant omissions and delays in taking the medication through alerts, promoting the medication adherence. However, this challenge is, by itself, a set of challenges when inserted in a project on Ambient Intelligence, because any kind of alert is very obtrusive. The annoyance is greater if there are needed medication alerts for multiple users in the same place. Since the target users for the proposed system in this work are the elderly, the biggest challenge lies in the interaction modes, because abilities change as people grow older, and the cognitive, physical and sensory functions attenuate over time [4].

Besides the older people's perception, motor and cognitive skills, the knowledge and their experience with new technologies and their difficult and limited availability to adapt and to learn must be also considered. This does not mean that interfaces for older people should be specialised but it means that this requires approaches which are outside the scope of typical user centred design in order to provide useful interactions [4].

Overall, this work focuses on the design of a multi-user alert system which will promote medication adherence contributing for an efficient solution to help the elderly take their medication independently. Another goal of this paper is to present a plan of research to deploy and evaluate the proposed solution, indicating the methodology which should be adopted and how the research results can be obtained.

2 System Overview

The main objective of this section is to describe clearly the system proposed, the projected architecture and its key functionalities.

2.1 System Description

The main concept of the proposed system is to provide each person whose medication must be managed with a device containing a vibrating alarm and a LED which notifies the user who has to take medication at that time.

This device, called tag, also determines the location of the tag carrier. Therefore, when a user is notified, they have to go to the pill cabinet to take the medication. The system controls this action verifying whether the user goes to the area of the pill cabinet after receiving notification. When the user does not go to the area of pill cabinet, the system keeps on notifying during a certain period. While the user is taking the medicine, they must press the button of the tag to confirm it explicitly. As the system is provided for multiple users, it is important to include in each tag the identification of its user, to avoid any mistakes.

In addition, if the system detects that the tag is not being carried by the user, the system must dynamically and proactively enable an alternative notification such as notification by telephone or, if not successful, inform the caregiver.

The alternative notification should be activated in two situations: if the user carries the tag but does not go to the area of pill cabinet and does not press the button to confirm the taking of medication; or if the tag is not carried by user.

The system must provide a back-office to set up the system. In the back-office the caregiver has to introduce the users of the system, specify the location of the pill cabinet in the map of the house and also set the hours of taking the medication for each of them. Introducing a user means creating a user and indicating the settings of utilisation of the system for that user, for example inserting the phone number for alternative notifications, choosing the mode of communication with the caregiver and the time required to activate the alternative notification. For this project it is not necessary to indicate what medicines to take but only the hours to take them, since this project is only focused on the alert function. Nevertheless, these considerations should be left open for future extensions and adaptations of the system.

2.2 System Architecture and Technologies

The proposed system is based on a RTLS (Real Time Location System). A RTLS is a fully automated system which provides the continuous monitoring of the location of assets or people on a campus area. It does this in real-time, delivering information to authorised users or agents via a corporate network through application software or application programming interfaces [5].

The backbone component of the projected system is a Wi-Fi network because all the system operates over it. It can operate any brand or generation of Wi-Fi network (802.11 a/b/g/n) while offering sub room-, room-, floor- and building-level accuracy. Given that, it is a Wi-Fi standards-based solution which requires no additional readers, new cabling, choke-points or exciters. It only uses existing Wi-Fi standard access points as the reference devices for tag location [5].

Another important component in this system is the tags. The tags are devices which are attached to assets or carried by people you want to control and track. The tags used by an RTLS are called Wi-Fi tags. There are different types and formats of Wi-Fi tags [6], but the type of tags chosen for this system is a wristband tag to make it easy for people to carry it.

The wristband tag is designed to be worn comfortably on a person's wrist or ankle. The tag is waterproof and therefore the tag wearer can comfortably take a bath or shower while wearing it. The tag includes a two-colour LED, a vibration alarm for alerting the tag wear and also has a customizable call button to send emergency or status messages [6]. The wristband tags provide a high degree of location accuracy ensuring precise and reliable location about a meter in both confined and open spaces [7, 8] and it only uses signal strength readings from a Wi-Fi network to build the location algorithm [9].

The tag emits a signal periodically to the network, which is at that time captured by the system to calculate location [6, 10]. The access points are the reference devices for locating tags [5], allowing to determine if the user goes to the location of pill cabinet after notification.

Furthermore, the tag has a highly sensitive motion sensor which detects motion and if the tag has been removed or tampered with [6, 10]. This function will be used to activate the alternative notifications in case the tag is not being carried.

The tag also has a button which can be configured to work as a help button or to signal any other message when pressed [6, 10]. In this system, this button will be used by the user to confirm that medication has been taken after notification.

The projected architecture for this system (represented in Fig. 1) has two agents which compose the main system functionalities: location agent, which is able to locate tags within the Wi-Fi network; and notification agent, which is responsible for issuing the alerts provided in the system.

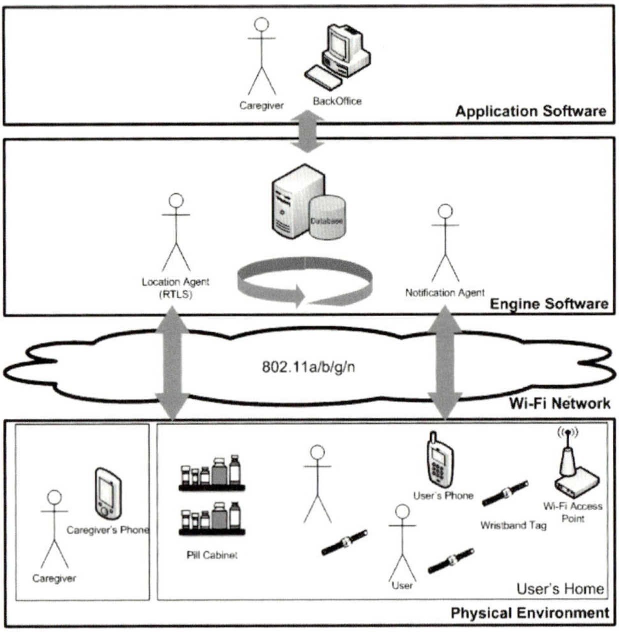

Fig. 1 System architecture

In addition to these, there is another important engine component – a database. In this solution, the database assumes the role of repository of critical data. All configuration settings are registered and stored here and it is the responsible for the data supplying of the other engine agents.

The component designated *Application Software* represents the system back-office available to the caregiver. It is not represented in *the Physical Environmen*t because it is not a system functionality but a system configuration functionality.

The main advantage of the proposed system is that it uses some type of user's Wi-Fi networking and it is able to take advantage of the two-way communication capabilities inherent to Wi-Fi [8]. As previously mentioned, the tags work in this system both as a sensor to determine if the user was in the cabinet after notification and as an actuator, since this is the main device to notify the users.

3 Research Objectives

As it is possible to realise by the analysis of the architecture presented above, there are a number of other objectives to be achieved to implement this solution [11, 12, 13]:

- **Acceptability and Usability**
  The final result should be accepted by potential users. A solution to be acceptable by its users should achieve the main goals with effectiveness, efficiency and satisfaction in the context of use [14].
  Thus, the solution should be designed in a way users can perform the tasks associated accurately and completely (effectiveness); can perform work quickly in the expected period of time (efficiency); and like the product in terms of ease of use and usefulness (satisfaction). A solution will become ubiquitous only if it has high acceptance by users.
  For acceptability and usability of this solution the following situations will be considered: provide interfaces for easy usability; make unambiguous interaction modes available; choose the less obtrusive equipment and interaction modes; and evaluate the acceptance and usability of the system.
- **Availability and Adaptavity**
  The system must do its job even in exceptional conditions and help proactively, if needed. During usage, the system must adjust its behaviour to user goals, user competence, and user interests.
  For availability and adaptability of this solution the following situations will be considered: adapt the behaviour of the system according to the user's situation; ensure that there is always a way to notify the user, ultimately by the caregiver; and guarantee that all algorithms are efficient and accurate to make the system available in real time.
- **Extensibility and adaptability**
  The system must be easy to update functions and data in order to suit the individual elderly user and must support its extension by new components at runtime, for example mechanisms which provide other functions of administration of medication.
  For extensibility and adaptability of this solution, it will be guaranteed that the solution is open. It means that developed algorithms, interfaces and APIs should be as flexible as possible to develop other functionalities and enable its integration based on current and future needs of users.
- **Robustness and Safety**
  The system must be extremely robust against all kinds of misuse and errors. Wrong inputs must not lead to a system malfunction or crash. The system must do exactly the job it was designed for and faulty system components and exceptions must never result in system misbehaviour.
  For robustness and safety of this solution the following situation will be considered: users should always be notified independently of the way even if it is detected which tags have been removed or tampered with.

- **Security**

  A living assistance system, although continuously monitoring people, must guarantee a well defined degree of privacy for those under observation. The privacy rules must be accurately formulated and verified.

  For security of this solution the following situations will be considered: ensure that location data about users is only used to confirm their location on the area of the pill cabinet after notification; and guarantee that equipment used in solution does not cause any interference with life critical devices such as pacemakers and other healthcare devices.

4 Methodology and Associated Work Plan

An user-centred design and research typically follows an iterative cycle which comprises some fundamental processes, in which we understand, study, design, build and evaluate the solution [15]. These five stages include very important and fundamental objectives and tasks to any research work. The previous sections of this paper have content which fits in the first three stages of the iterative cycle. Thus, the last two stages are approached in this section.

The methodology intended for the fourth stage (building stage) of this solution consists in the construction of prototypes. The prototypes are used as vehicles to provide a grasp of how future implementations can look like regardless of their immediate feasibility [16]. Bonner and Van Schaik [17] use the terms low- and high-level and define a high-fidelity or high-level prototype as one "where all or most of the functionality and often the form of the interface is fully represented". Older people do not see low-fidelity prototypes as representing an application, but they work well with high-fidelity prototypes [4, 18]. For these reasons, the best type to prototype the system in this context is a high-fidelity prototype. The first phase of the prototype development is the deployment of the solution in a house where more than one potential elderly user resides over a period of three months.

After evaluating the first prototype, the solution will be gradually deployed in slightly different contexts, i.e. in environments with users with different cognitive and motor capabilities. The main intent is to determine the type of users to become user target of the solution.

The evaluation should be made in the prototype deployments, because it is easier to understand the impact of solution on the real environment and context than in labs. Nonetheless, before the first deployment, in the early stages some lab studies should be made in order to evaluate the technological feasibility.

The evaluation is necessary and very important to determine if the research objectives are achieved and completed. For this system, the evaluation can be done getting the user opinion by interview or observation because these methods reflect the system usefulness and the level of user satisfaction. However, if this information is insufficient or inconsistent, cooperative evaluation can be used to validate the guidelines and collect any additional problems [19].

It is also significant to have some quantitative data for evaluation as example of error rate when using the system. To obtain these data, it is necessary to create mechanisms in the prototype systems, which count the successful and unsuccessful situations during the period of test. Another important aspect in the system evaluation is to compare results in the different groups of users of the various systems prototypes in order to determine the potential users.

The evaluation of the back-office system can be made heuristically. "Heuristic evaluation is a usability engineering method for finding the usability problems in a user interface" [20]. Hence the Jakob Nielsen's ten heuristics can be used to do it.

5 Final Considerations

Providing mechanisms to reduce the significant omissions and delays in taking the medication through alerts, promoting the medication adherence, has been a challenge in ubiquitous computing.

The solution presented in this paper is a contribution for reminding medication, especially in multi-user environments. Moreover, the architecture proposed for this solution is consistent with the main research objectives.

In addition, another great advantage of this solution is that it provides extensibility. The system can be updated and upgraded with more functions and data in order to suit the needs of individual elderly users as well as with new components which provide other functions of administration, for example providing the user with information and/or instructions about the medicine; adding pill cabinet which automatically dispenses medication; logging date and time when medication is taken and report it to the caregiver; and reporting the system state (e.g. charge level of battery tags) to the caregiver.

References

1. Technologies to Help Older Adults Maintain Independence: Advancing Technology Adoption. Center for Technology and Aging (2009), http://www.techandaging.org/briefingpaper.pdf (accessed January 8, 2010)
2. Patient Compliance Medication Adherence Statistics & References (2009), http://www.epill.com/statistics.html (accessed December 27, 2009)
3. Fishkin, K., Wang, M.: A Flexible, Low-Overhead Ubiquitous System for Medication Monitoring. Intel Research Technical Report IRS-TR-03-011 (2003)
4. Hawthorn, D.: Interface design and engagement with older people. Behav. Inf. Technol. 26(4), 333–341 (2007)
5. Ekahau, Real Time Location System (RTLS) Overview. Ekahau (2009), http://www.ekahau.com/products/real-time-location-system/overview.html (accessed January 8, 2010)
6. Ekahau, Ekahau T301 Tags. Ekahau (2009), http://www.ekahau.com/products/real-time-location-system/wi-fi-tags.html (accessed 8, January 2010)
7. Hallett, T.: WiFi software tracks you down (2002), http://news.zdnet.com/2100-1009_22-125870.html (accessed January 8, 2010)

8. Ekahau, Wristband Tag for Wi-Fi-based Real Time Location Tracking. Ekahau (2009), http://www.ekahau.com/news/readallnews/press-releases/187-ekahau-introduces-wristband-tag-for-wi-fi-based-real-time-location-tracking-.html (accessed January 8, 2010)
9. Ekahau, The Science of High-Performance Wi-Fi Real Time Location System. Ekahau (2009), http://www.ekahau.com/products/real-time-location-system/overview/technology.html (accessed January 8, 2010)
10. Ekahau, Solution Sheet Patient Safety. Ekahau (2009), http://www.ekahau.com/images/stories/solution_sheet_patient_safety_letter.pdf (accessed January 8, 2010)
11. Maciuszek, D., Aberg, J., Shahmehri, N.: Evaluation and Refinement of a Design Framework for Generating Dependable Virtual Companions for Later Life. In: Proceedings of the 3rd International Conference on Smart Homes and Health Telematics (ICOST 2005), Sherbrooke, Canada, pp. 50–64 (2005)
12. Maciuszek, D., Shahmehri, N., Aberg, J.: Dependability requirements to aid the design of virtual companions for later life. In: Proceedings of the First HEAT: The Home and Electronic Assistive Technology Workshop, York, UK, pp. 51–60
13. Nehmer, J., et al.: Living assistance systems: an ambient intelligence approach. In: Proceedings of the 28th international conference on Software engineering, pp. 43–50. ACM, Shanghai (2006)
14. ISO 9241-11 Guidance on Usability (1998)
15. Harper, R., et al.: Being Human: Human-Computer Interaction in the Year 2020, ed. M. Research (2008)
16. Michahelles, F., Schiele, B.: Experiencing Technology before it exists: A Case Study. In: A Pervasive 2005 Workshop - What makes for good application-led research in ubiquitous computing?, Munich, Germany (2005)
17. Bonner, J., Schaik, P.V.: The use of high and low level prototyping methods for product user interfaces. Contemporary Ergnomics 1998, ed. M.H., London (2004)
18. Blackler, A.: Applications of high and low fidelity prototypes in researching intuitive interaction. In: Proceedings of Design Research Society Biennial Conference, Sheffield, United Kingdom, pp. 154–170 (2008)
19. Zaphiris, P., Pfeil, U., Xhixho, D.: User Evaluation of Age-Centred Web Design Guidelines. In: Proceedings of the 5th International Conference on Universal Access in Human-Computer Interaction. Addressing Diversity. Part I: Held as Part of HCI International 2009, pp. 677–686. Springer, San Diego (2009)
20. Nielsen, J.: Heuristic Evaluation (2005), http://www.useit.com/papers/heuristic/ (accessed January 8, 2010)

Ontology and SWRL-Based Learning Model for Home Automation Controlling

Pablo A. Valiente-Rocha and Adolfo Lozano-Tello

Abstract. In the present paper we describe IntelliDomo's learning model, an ontology-based expert system able to control a home automation system and to learn user's behaviour. This system obtains its own data from the user's daily activities. From these data, IntelliDomo can detect behaviour periodic patterns and it can be able to generate SWRL rules that anticipate the user's day-to-day activities. One of the main objectives IntelliDomo seeks is to be as transparent as possible for the user. The user has nothing to configure, the system learns his/her habits and anticipates his/her actions.

1 Introduction

The use of Ambient Intelligence (AmI) is one of the areas which are rapidly gaining importance in the application of intelligent systems in companies and homes.

One of the main characteristics AmI systems seek is the gift of interaction with the user so that communication is as natural as possible and the configuration of preferences takes the shortest time for the user. Ideally, the user would have to finally forget the presence of the system by getting used to home automation devices anticipating to his/her needs and preferences.

A correct way of representing home automation systems and behaving rules is throughout the use of ontologies and production rules based on the concepts established on these ontologies.

For many years, ontologies have generally become the way of representing the systems upon current knowledge [1]. This is due to the fact that they are designed in order to make their knowledge easily reusable and shared by communities and users of the same domain. Besides, the acceptance of OWL (Ontology Web Language) as a de facto language for its representation and the development of tools such as Protégé for its construction have favored its wide use in many fields, especially in the Semantic Web [2].

Pablo A. Valiente-Rocha · Adolfo Lozano-Tello
Universidad de Extremadura. Quercus Software Engineering Group
Escuela Politécnica, Campus Universitario s/n, 10071, Cáceres, Spain
e-mail: {pvaliente,alozano}@unex.es

J.C. Augusto et al. (Eds.): ISAmI 2010, AISC 72, pp. 79–86.

As far as the representation of production rules is concerned, SWRL (Semantic Web Rule Language) [3] has been adopted by W3C as the representing standard for production ontology-based rules. SWRL is a very complete language based on OWL; moreover, there are several plug-ins for the editing of rules in the Protégé tool, which have favored its spreading and its use. There are currently some reasoners based on OWL and SWRL that allow maintaining consistency and inferring values for the attributes of classes in the ontologies.

The use of ontologies as representing bases, together with the representation of rules in SWRL in AmI systems, will provide the most precise definition of the taxonomy of physical devices that may exist in a system, the attributes of these devices and the possible relations among them. Furthermore, these representations will be more reusable by other users and will favor the completion of the classification of home automation components, the useful information to be represented upon these components and the connection rules that will allow deducting new information regarding the values of other components.

The present paper describes IntelliDomo's learning model, an ontology-based AmI system for the control of home automation systems and the learning of the users' habits. In section 2, we identify some publications using ontologies in home automation systems and another two that propose some learning model for an AmI system. In section 3, we describe IntelliDomo's learning system, and in its subsections, we describe each of the components from the model software; and we also present a laboratory example to show the use of the model. Finally, section 4 is dedicated to conclusions and future lines.

2 Existent Works about the Use of Ontologies on AmI Systems

In recent years a lot of initiatives for AmI systems controlling and representing are emerging from both research team and domotic companies [4], which in most cases try to establish their own standard. However, the amount of literature about intelligent domotic environments, which makes use of ontologies as representing bases, is not very extensive.

DomoML is one of the main reference works [5, 6]; it is a markup language focused on defining a communication method among domotic devices. Three ontologies are proposed -DomoML-Env, DomoML-Fun and DomoML-Core- as a base to build an architecture that allows setting up hierarchies and position and categorize domotic devices. DomoML operates as a communication protocol among devices and as a representation language to describe resources and functionalities.

Related to the previous project, DogOnt [7] proposes a system able to recognize the devices that comprise the domotic environment automatically using both, DogOnt Ontology in domotic devices; and DogONT Rules, which allows the association of intalled devices.

ReWiRe project [8] is a framework that supports networked heterogeneous environments and allows the environment configuration to change over time. ReWire ontology provides with the semantics required to create a system that can rewire during usage.

CASAS [9] shows us some techniques so interesting to get automation of sequential complex activities that adapts to the user's preference. CASAS collects different kind of information that helps one to better model the complex environment, and also show us a way to find and discover frequent and periodic patterns of activity.

MAV [10] is designed to learn a model of the inhabitant of the intelligent environment, automate devices and adapt it over time to accommodate shifts and drifts in the inhabitant's life patterns.

Some of the previous projects use ontologies in order to represent home automation knowledge, others control their operations with rules, and others handle different techniques for discovering patterns about inhabitants' behaviour. IntelliDomo matches up some of the above objectives, using behaviour detection techniques to automatically produce SWRL rules. These rules try to anticipate the behaviour in user's daily activities.

3 IntelliDomo's Learning Model

IntelliDomo is an expert system able to control the components of a home automation system automatically and in real time using SWRL rules. The overview model was presented at [11] and explained in more detail at [12]. In this paper, we present IntelliDomo's learning model. This plug-in generates rules automatically depending on the user's previous behaviour. It is in charge of the learning of the system and is currently in an advanced development stage. This approach is similar to aCapella [13] ones, because it supports end-users in programming by demonstration recognition-based context-aware behaviors.

IntelliDomo is able to use behaviour SWRL rules that have been designed in the system. The aim is to detect behaviour patterns regarding the rule execution frequency in order to be able to anticipate to daily user's tasks.

Section 3.1 describes with more detail the rule management system. Afterwards, Section 3.2 studies the learning module that is used in IntelliDomo.

3.1 SWRL Rules Management

The use IntelliDomo makes of SWRL rules is not aimed to check the information consistency or to complete its ontology, but to establish the behaviour model that will guide the actions of the system in real time.

SWRL rules are initially stored in a database that provides persistent storage to the system. Once the SWRL rules are stored in that database, they can be modified without running IntelliDomo. Each SWRL rule has its own unique priority used to organize the system behaviour. This avoids two SWRL rules, that propose opposite actions, being fired at the same time producing wrong system behaviour. Two SWRL rules can propose opposite actions depending on the problem they are trying to solve. For instance, the consequent of Rule-A could propose: "raise the shutter", and the consequent of Rule-B: "lower the shutter". The antecedent of each SWRL rule could be either true because Rule-A could be trying to find the adequate temperature (Comfort) and Rule-B could be trying to minimize the possibility of an intrusion (Security). The priority assigned to each SWRL rule will determine which rule should be fired first to resolve the conflict set.

To be able to use each one of these SWRL Rules, IntelliDomo will load them into the ontology and store them as OWLIndividuals, keeping a two-way sync between SWRL Rules stored on the ontology and those stored on the database.

name	rule	username	priority	frec
Raise Shutter	Temperature(?x) ∋ value(?x, ?y) ∋ swrlb:greaterThan(?y, 25) ∋ value(roomShutter, 0) ⟶ value(roomShutter, 10)	default	6	105
Lower Shutter	Temperature(?x) ∋ value(?x, ?y) ∋ swrlb:lessThan(?y, 20) ∋ value(roomShutter, 10) ⟶ value(roomShutter, 0)	default	7	104

Fig. 1 This table shows two possible SWRL rules from IntelliDomo's database. The name, the syntax, the owner and the rule's priority are shown.

The system stores the actions carried out after each rule execution as well as the set of facts that have led to the execution of such rule on a log table. Time and date of event is also stored. An event is considered to be an action that takes place at a home automation system and causes a change in the current values of the physical devices. An event can be activated in two ways: Activated by the user (for example turning on the light) or activated by Jess [14] inference engine firing a SWRL rule. Thus, the fact of "raise the shutter" is considered an event that can be started by the user or automatically launched by IntelliDomo.

An event is always stored next to its start date and time, its set of facts (antecedent of the rule) and the name of the rule that has started such event. If the user has started the event (see figure 3), then, the name of the rule will be empty and the values that have been modified will appear on the list of facts field, i.e. the actions that the user has carried out.

ruleName	date	facts
RaiseShutter	2009-12-10 19:59:51.981	ST1.value=30;ST1 is Temperature(sensor); roomShutter.value=0;

Fig. 2 This table shows a row from the event log table. IntelliDomo has fired this rule.

Having registered the date and time of each event, IntelliDomo can look for frequent and periodic patterns of activity and obtain adequate information about the user's habits.

3.2 Learning Module

Many factors have been taking into account for the learning task when deciding which information was useful and which was of no use. The learning module is responsible for two main tasks: Finding new periodic patterns on the events that occur over time (see section 3.2.1) and Adding/Removing IntelliDomo's SWRL rules based on these patterns (see section 3.2.2). The aim is to learn the most common tasks performed by the user and to anticipate to the user's habits. The way to carry out this learning involves periodically checking the tasks performed by the user and finding out behaviour daily patterns without getting involved in his/her day-to-day activities.

In the next section, we describe what methods are used by IntelliDomo to study the user's behaviour. And in the subsequent section we detail the SWRL rule auto-generation system.

3.2.1 Finding Periodic Patterns

IntelliDomo stores log registries from each actions or events that take place in the system. As described before, these events can be launched by the user or by any change over the devices which have made one SWRL rule be fired. Each log entry is associated to the date and time when such event happened. Using this information, the learning module analyzes the actions occurred in the system and tries to find a recurrence pattern with an acceptable error rate. The aim is to discover the most repetitive routines carried out daily by the user and manage the system to anticipate to those actions.

IntelliDomo checks some degree of periodicity: daily, weekly, monthly, and annual. Daily periodicity is divided into weekends and weekdays. IntelliDomo collects information for at least one month to detect behavior periodic patterns. The more amount of information IntelliDomo has, the more precise its deductions will be. When an acceptable rate of periodicity is detected in an event, IntelliDomo will automatically build the necessary SWRL rules so that they can anticipate to the actions taking place in such event at the right day and time.

Frequent user's events would not regularly happen at the same minute due to human beings' erratic nature. Therefore, a time range that determines whether two actions have occurred simultaneously is established. IntelliDomo uses an uncertainty range of ten minutes by default. Nevertheless, this value can be modified in the setting of the learning module.

Furthermore, there is no guarantee that a user would perform the same actions every day of the week. Therefore, we also use the parameter "frequency error" to ease pattern discovery and to consider if an event is regular. By default, IntelliDomo thinks an event is regular if it occurs more than 90% of the time. This parameter can also be modified.

One of the tests that we have performed in our laboratory threw these results:

The user arrives every day at work between 8:00 am and 8:09 am. The first thing he/she does is "To turn on the light" and "To raise the shutter". These actions are performed daily excluding weekends and holidays. When the user finishes his/her workday between 1:40 pm and 3:00 pm, he/she comes back home and "Turn off the light" and "Lower the shutter".

Although throughout the week more events have occurred during the course of days, we are only focusing on the ones related before to show a brief view of the table log.

ruleName	date	facts
	2009-12-07 08:01:51.981	officeLight.value=1;officeLight is Light; officeShutter.value=10;officeShutter is Actuator;
	2009-12-07 14:59:52.956	officeLight.value=0;officeLight is a Light; officeShutter.value=0; officeShutter is Actuator;
	2009-12-08 08:05:30.681	officeLight.value=1;officeLight is Light; officeShutter.value=10;officeShutter is Actuator;
	2009-12-08 14:57:46.152	officeLight.value=0;officeLight is a Light; officeShutter.value=0; officeShutter is Actuator;

Fig. 3 Example showing what happens when the user goes into his/her office during the week

With this information, IntelliDomo detects that some patterns are repeated during the workdays. These patterns can be useful to build a new SWRL rule and to anticipate to the user's actions. Firstly, the pattern "officeLight=1", secondly the pattern "officeLight=2", thirdly the pattern "officeShutter=10" and finally the pattern "officeShutter=0".

To detect these periodic patterns on the data sequence, IntelliDomo's learning module uses a modified version of Agrawal's Apriori algorithm [15]. Our version has been simplified and adapted to detect the sequence of patterns inside IntelliDomo's log tables. With this information, IntelliDomo adds or deletes the appropriate SWRL rules, as is described in the next section.

3.2.2 Adding/Removing SWRL Rules Automatically

Once the behaviour pattern has been obtained and the time and the action of the event are known, IntelliDomo has all the necessary information to build a SWRL rule with a simple and valid syntax. These SWRL rules, will be added to IntelliDomo's rule set and they will be assigned the lower priority based on the other ones rules in the database. This is due to the rules in the database have been directly designed by the user to control his/her home automation system and he/she has defined the desired priority to establish the importance on each rule. The owner of the SWRL rules generated by the learning module will be "idomo". Thus, we will be able to discriminate between the rules that have been automatically created from the ones that have been built by the user.

SWRL rules are built with the help of IntelliDomo's parser, which is able to convert the logs generated by the learning model into SWRLAtoms. These SWRLAtoms will have this appearance: Property(Instance, value) as it is shown in figure 4. Later, IntelliDomo will update its ontology and the corresponding tables from the database to incorporate these new SWRL rules. Each SWRL rule has its own timestamp. Thus, the module of the system responsible for running the inference engine will only be able to fire the rules within the current time range. With this in mind, if the system fires one rule and the user carries out the opposite action several times within temporal uncertainty range, the system might automatically delete that rule, because it would have detected that the user did not want that action to take place. The system would only delete one rule if it belongs to "idomo".

For instance, when IntelliDomo detects that it is a few minutes before 8:00 am, it will fire both rules "officeLight1" and "officeShutter10" (these patterns have been previously learned). If the user lowers the shutter during the next ten minutes for several days, IntelliDomo will remove the rule "officeShutter10" to memorize again when it should or should not raise the shutter.

ruleName	date	Rule
officeLight1	2010-01-07 07:58:00.981	Light(officeLight) ∍ value(officeLight, 0) value(officeLight, 1)
officeLight0	2010-01-07 14:58:00.956	Light(officeLight) ∍ value(officeLight, 1) value(officeLight, 0)

Fig. 4 Example showing SWRL rules created by IntelliDomo

4 Conclusions and Future Lines

In the present paper, we have presented IntelliDomo's learning model, a SWRL and Pattern Discovery-based model. The main goal is trying to learn the user's habits and to anticipate to his/her needs. With the help of the ontologies, we can make a more precise taxonomy of home automation devices and categorize the learned knowledge for reusing it as needed.

We have completed different tests in our laboratory, and we have verified that IntelliDomo is able to learn the user's habits in a single-user environment. Furthermore, it can anticipate to his / her activities. However, we must test IntelliDomo's learning model in real installations and evaluate the suitability of the proposed rules. Therefore, in the coming months, we will be checking the module in several particular houses.

For the future work, we are inclined to provide IntelliDomo with the ability to recognize more than one user at a time. Currently, IntelliDomo has the necessary data structures to allow interaction with more than one user, but we are still working in recognition modules to carry out this task.

Acknowledgement. This work has been developed under support of Junta de Extremadura Project (PDT08A023).

References

1. Fensel, D.: Ontologies: Silver bullet for knowledge management and electronic commerce. Springer, Berlin (2001)
2. Berners-Lee, T., Hendler, J., Lassila, O.: The semantic web. Scientific American, 29–37 (2001)
3. Horrocks, I., Patel-Schneider, P., Boley, H., et al.: SWRL: A semantic web rule language combining OWL and RuleML. (2004), http://www.w3.org/submission/2004/SUBM-SWRL-20040521/

4. Gasson, M., Warwick, K.: D12.2: Study on emerging AmI technologies, FIDIS Deliverables 12 (2007)
5. Sommaruga, L., Perri, A., Furfari, F.: DomoML-env: an ontology for human home interaction. In: Procs. of Semantic Web Applications and Perspectives, Trento, Italy, December 14-16 (2005)
6. Furfari, F., Sommaruga, L., Soria, C., et al.: DomoML: The definition of a standard markup for interoperability of human home interactions. In: Procs. of the 2nd European Union Symposium on Ambient Intelligence, pp. 41–44 (2004)
7. Bonino, D., Corno, F.: Dogont-ontology modeling for intelligent domotic environments. In: Sheth, A.P., Staab, S., Dean, M., Paolucci, M., Maynard, D., Finin, T., Thirunarayan, K. (eds.) ISWC 2008. LNCS, vol. 5318, pp. 790–803. Springer, Heidelberg (2008)
8. Vanderhulst, G., Luyten, K., Coninx, K.: ReWiRe: creating interactive pervasive systems that cope with changing environments by Rewiring. In: 4th Int. Conf. on Intelligent Environments (IE 2008), Seattle, July 21-22, pp. 1–8 (2008)
9. Rashidi, P., Cook, J.: Keeping the resident in the loop: adapting the smart home to the user. IEEE Transactions on systems, man, and cybernetics - part A: systems and humans 39(5), 949–959 (2009)
10. Youngblood, G., Cook, D., Holder, L.: Managing adaptive versatile environments. Pervasive and Mobile Computing 1(4), 373–403 (2005)
11. Valiente-Rocha, P., Lozano-Tello, A.: Control model of domotic systems based on ontologies. In: 2nd Int. Conference on Agents and Artificial Intelligence, Valencia, Spain, January 22-24, pp. 470–473 (2010)
12. Valiente-Rocha, P., Lozano-Tello, A.: Ontology-based expert system for home automation controlling. It will be presented at The 23rd International Conference on Industrial, Engineering & Other Applications of Applied Intelligent Systems, Córdoba, Spain (June 2010)
13. Dey, A., Hamid, R., Beckmann, C., et al.: A CAPpella: programming by demonstration of context-aware applications. In: Conference on Human Factors in Computing Systems, Vienna, Austria, April 24-29, pp. 33–40 (2004)
14. Friedman-Hill, E.: Jess In Action. Manning Press, Greenwhich (2003)
15. Agrawal, R., Srikant, R.: Mining Secuential Patterns. In: ICDE 1995, pp. 3–14 (1995)

Model-Driven Development for Ubiquitous MAS

Jorge Agüero, Miguel Rebollo, Carlos Carrascosa, and Vicente Julián

Abstract. This paper presents an approach to design and deploy Multi-Agent Systems (MAS) using MDD (Model-Driven Development), in a fast and simple way. This approach is specifically addressed to systems including agents that must be executed on mobile or embedded devices. The user will design the system for different platforms by means of unified agent models (using UML-Like models). There will exist different automatic transformations to obtain the specific code for different target platforms from these unified models. On this way, the deployment process of ubiquitous applications (with mobile/embedded agent) is simplified.

1 Introduction

Ubiquitous computation is a paradigm in which technology is virtually invisible in our environment, improving life's quality[7]. Environments with ubiquitous computation are very common nowadays, due to the appearance of *new objects* (of daily usage) with different technological capabilities, because they incorporate different electronic devices. So, it is easy to think that this paradigm implementation requires, from a designer point of view, the development of applications in different software and hardware platforms depending on the diversity of the objects in the environment. This raises big challenges. MDD is a technique allowing to obtain code for different execution platforms from models by means of consecutive transformations. This methodology can be applied in the development of embedded agents for ubiquitous computation, where different technologies and developing platforms coexist (nowdays, one way to implement ubiquitous computation is with embedded intelligent agents). The agent (software) that runs on embedded hardware (computer) is

Jorge Agüero · Miguel Rebollo · Carlos Carrascosa · Vicente Julián
Departamento de Sistemas Informáticos y Computación
Universidad Politécnica de Valencia
Camino de Vera S/N 46022 Valencia, Spain
e-mail: {jaguero,mrebollo,carrasco,vinglada}@dsic.upv.es

J.C. Augusto et al. (Eds.): ISAmI 2010, AISC 72, pp. 87–95.

usually called embedded agent[8]. Each embedded agent is an autonomous entity allowing to cooperate with other agents.

This work presents a method to obtain embedded agents (code templates) to be executed in different mobile platforms applying the basic ideas of MDD. That is, to design ubiquitous applications with agents using models or abstract concepts independently of the implementation details and any platform detail. After that, the platform specific embedded agent is generated by automatic transformations. In this way, a non-expert programmer will be able to develop systems with ubiquitous agents, reducing the gap between design and implementation. To check the viability of the proposal, the obtained executable code is implemented in two mobile agent platforms, ANDROMEDA[1] and JADE-Leap[4].

The document is structured as follows. Section 2 shows how to apply the MDD approach to develop ubiquitous agents according to the proposal presented in this paper. Next, section 3 explains through an example how to implement ubiquitous agents using MDD for different platforms. Conclusions are presented in section 4.

2 Using MDD to Develop Embedded Agents

MDD approach uses and creates different models[10]. Models can be classified in three groups depending on its abstraction level: Computation Independent Model (CIM), that details the general concepts independently if they are going to be implemented by a machine or not; Platform Independent Model (PIM), which represents the system functionality without consider the final implementation platform; and the Platform Specific Model (PSM), obtained by combining the PIM with specific details of the selected platform.

Only a few agent development methodologies have integrated the MDD techniques in the MAS design. The most relevant are: INGENIAS[11], TROPOS[12] and PIM4AGENT[9], AML[6] and AUML [3]. These are proposal to apply MDD to the agent modeling process, however AML and AUML does not arrive to implementation details. TROPOS and INGENIAS proposals are based on specific models of their methodology. PIM4AGENTS do not modeling the Virtual Organizations.

Now, design agent-oriented systems using MDD, is to generate agent code on different platforms. That is, to concentrate the development of the application from an unified agent model and to apply different transformations to get implementations for different platforms. Figure 1 illustrates how it is possible to obtain code to different agent platforms from unified models. In a first phase, as before commented, this work has been focused to generate code for two agent platforms: JADE-Leap[1], which is one of the most known agent platforms for mobile devices, and the ANDROMEDA[2] platform, specifically designed to execute agents over the Android[3] OS.

[1] http://jade.tilab.com

[2] http://users.dsic.upv.es/grupos/ia/sma/tools/Andromeda

[3] Android System, http://code.google.com/android/

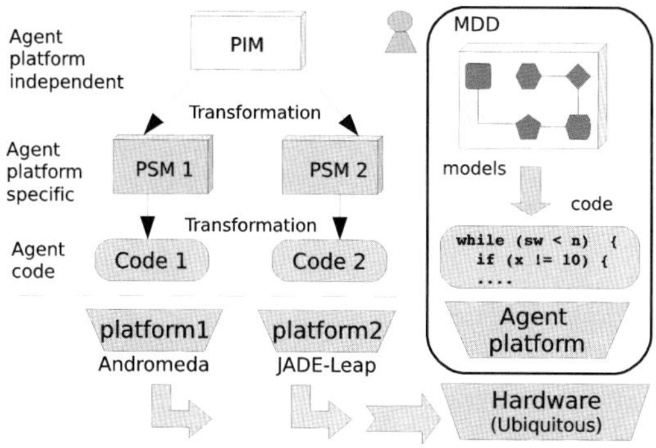

Fig. 1 MDD for different platforms

- **Agent meta-model**

An *Agent* is the basic entity of a MAS. So, at a high level of abstraction it is necessary to design an agent without taking into account specific details of concrete agent platforms. To do this an appropriated meta-model is needed. The proposed agent meta-model, called *agent-π (agent Platform Independent)*, is a set of interrelated components, each serving a specific function for the agent definition. The main components are: *Behaviours*, *Capabilities*, and *Tasks*; Table 1 summarizes the main components and another concepts employed in the agent meta-model. However, for a more detailed explanation of the model refer to [1].

Table 1 Main concepts employed in the agent-π

agent-π concepts	Description
Agent	The entity *agent* represented in any methodology. A rational and autonomous entity.
Behaviour	It encapsulates a set of capabilities activated; it represents the abstract concept of role.
Capability	It represents an event-driven approach to solve a specific problem.
Task	The know-how related to a specific problem.
BeliefContainer	An abstraction employed to represent the agent knowledge.
Goal	A specification of a state that agents are trying to achieve.
Message	The typical mechanism employed for intercommunication among agents.

- **Development process**

The agent design process is formed by a set of steps (mainly transformations) that will obtain the executable code. In order to do these steps a set of tools, which support the process, are required. The steps and tools employed at each design stage can be summarized as:

Step 1: to create diagrams (through graphical tools) which model the different behaviors, tasks, interactions, etc. of the agents. To perform this step, the Eclipse IDE with a set of *plugins* is employed. This plug-ins are mainly *EMF*, *Ecore*, *GMF* and *GEF*, which allow the user to draw the models that represent the agent. Obviously, the needed agent meta-model (*agent-π*) has been previously implemented and loaded into the development environment in order to generate the appropriated agent models.

Step 2: to select which platform the user wants to execute the agent. This phase corresponds with the PSM model definition of each agent. To do this, it is necessary to apply a model-to-model transformation (PIM-to-PSM). This is done using the Eclipse IDE and the ATL *plug-in* incorporating the appropriated set of transformation rules. It is important to remark that the same agent model can be transformed into different specific agent platforms. Figure 2(a) and Table 2 illustrates the component transformations between two agent meta-models (from *agent-π* to JADE-Leap). The agent concepts or components are mapped (or moved) from source models to target models. The subset of the transformation rules needed for ANDROMEDA platform are explained in detail in [2].

Table 2 Main rules of transformation between *agent-π* and JADE-Leap meta-models

Rule	Concept	Transformation
1	Agent	agent-π.Agent $\Rightarrow$ JADE.Agent
2	Behaviour	agent-π.Behaviour $\Rightarrow$ JADE.ParallelBehaviour
3	Capability	agent-π.Capability $\Rightarrow$ JADE.OneShotBehaviour
4	Task	agent-π.Task $\Rightarrow$ JADE.Behaviour
5	Events	agent-π.Event $\Rightarrow$ JADE.ACLMessage
6	Beliefs	agent-π.BeliefContainer $\Rightarrow$ JADE.Schema
7	Goal	agent-π.Goal $\Rightarrow$ JADE.Ontology
8	Message	agent-π.Message $\Rightarrow$ JADE.ACLMessage

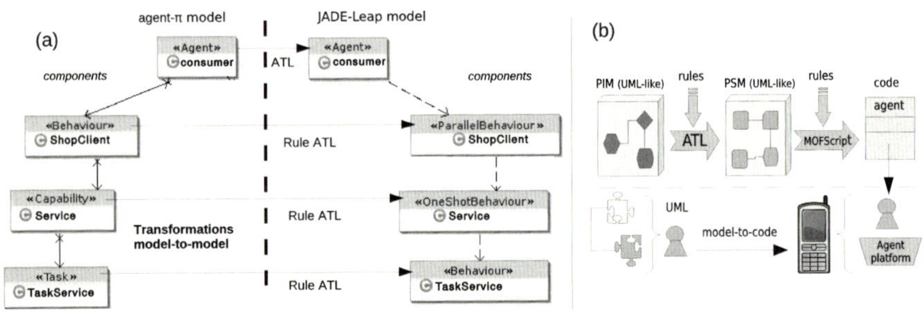

Fig. 2 ATL transformations between agent models(a) and overall process(b)

Step 3: to apply a transformation to convert the model into the agent code. To do this, we must use a PSM-to-code transformation. In this case we use *MOFScript* which is an Eclipse plug-in that uses templates to do the translation. These templates has been developed, as previously commented for JADE-Leap and ANDROMEDA. All the above steps are illustrated in Figure 2(b). The Figure shows how the agent can be converted from its model to its code using the MDD approach.

3 Implementing an Application Example

In order to illustrate this approach, a case study for an intelligent Mall is used. The intelligent Mall allows mobile users to know the product offered for a Mall based on the user profile(preferences), i.e., the user receives recommendations for possible purchase and use of Mall's Services. The Mall example is an application that facilitates the interconnection between clients (individuals, consumers) and providers (shops, Cinemas, Fast food); delimiting services that each one can request or offer.

However, before describing the implementation of Mall example, obviously some agent platform is needed. The Mall can be designed as a virtual organizations supporting an open agent society. The organization is implemented with the **THOMAS** platform[5], which allows the development of open MAS. Now, the agents can enter and leave of the organization consuming the services and products of the Mall. These agents are embedded on a mobile phone, using **ANDROMEDA**[1] or **JADE-Leap**[4] platforms.

- **Implementing the embedded agents**

This case study is modeled as an organization (Mall) inside which there are different organizational units (shops) that represent groups of agents. Each unit is dedicated to offer specific services and recommend products (due to space constraints, this example is briefly described). The Mall provides wireless data services, which allows mobile devices (mobile phones, netbooks, personal digital assistants) communicate with different Shops. This communication system is built upon a number of wireless standards, (by examples: Wifi, Bluetooth).

The Mall architecture is composed of many agents running on different platforms (see Figure 3(a)). The main agents identified in the systems are: (i) *Client Agent* is an user agent running on mobile devices (using ANDROMEDA and JADE-Leap platforms) that acts as a Personal Assistant (PA); (ii) *Shop Agent* consists of the shop agents, which act as access points or gateways for supply of services and products to mobile users; (iii) the *Mall* is a virtual organization which acts as a "container" of different shops (using the THOMAS platform). However, its special function is to allow the creation and registration of shops and their services. In addition of controls and synchronizes the information system. Figure 3(b) shows the interaction and dialog process among agent platforms used in the Mall example.

The agent implementation starts by defining the different components used in the agent. These components are part of the *agent-π* meta-model. In this case, we only describe agents running on mobile phones using ANDROMEDA or JADE-Leap platforms, due to space limitations of article. This *client* agent model has various behaviors that allow managing services offered by the shops. For reason of brevity the paper only explains the two most important behaviors of the client agent.

These two most important behaviors allow the agent to acquire the role of client and to discover the shops and send profiles. In Figure 4(a) these two behaviors can be observed. *Behaviour* `ShopDiscover` allow the agent to perform the activities necessary to find a shop and send the user profiles, while the other *Behaviour* `ShopClient` does the necessary tasks to consume and view the offered products

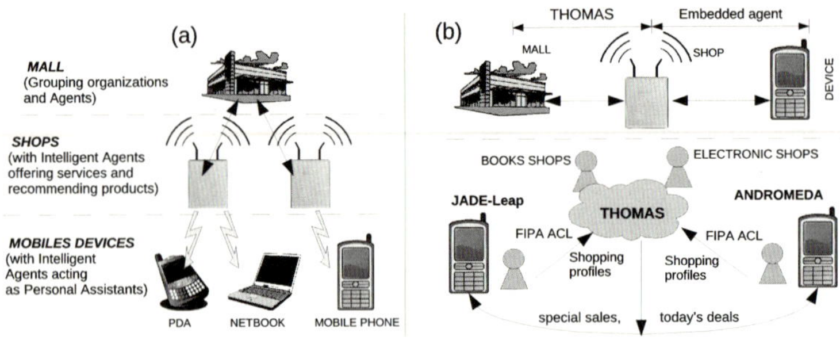

Fig. 3 The Mall architecture(a) and agent platforms(b)

and services. These behaviors are composed of two individual capabilities (see Figure 4(a)). As an example the *Behaviour* `ShopDiscover` includes the *Capabilities* `NetManager` and `Broadcast`. These *Capabilities* realize the actions of looking for a shop and send the user profiles respectively. The *Behaviour* `ShopClient` can be designed in a similar way, as can be seen in Figure 4(a).

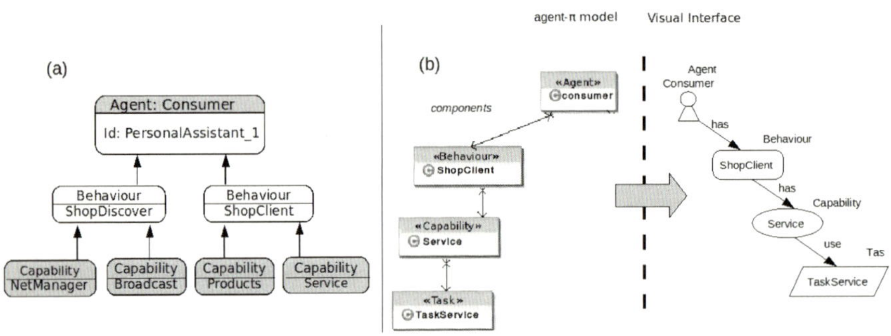

Fig. 4 Embedded agent model(a) and design in Toolkit(b)

Once we know the different components of the system, the design continues with the creation of embedded agents and the Mall system, which is modelled as an organization that will be executed over the THOMAS platform (how is modelled this organization is out of the scope of this paper, for more information please refer to [5]). This step is accomplished using the **EMFGormas Toolkit**[4] that allows users to "draw"/design (UML-Like) all the agent components according to the *agent-π* meta-model (first step, section 2). Once the agent model is designed, EMFGormas Toolkit allows verifying the designed system (in a semantic and syntactic way). Figure 4(b) shows a snapshot of the components of the client agent model and his correspondence with the *agent-π* meta-model.

[4] http://users.dsic.upv.es/grupos/ia/sma/tools/EMFGormas

Now, the next step is to generate agent code using this toolkit (second step explained in section 2). The designer must decide in which mobile platform will be executed each agent. In this case, the *client* agent is selected to be executed over an agent platform: ANDROMEDA or JADE-Leap. The **ANDROMEDA** platform has been designed using the same concepts of the *agent-π* model. So, an agent in ANDROMEDA is implemented using the same concepts employed in the abstract model. This design greatly simplifies the automatic transformation between PIM and PSM, because the ANDROMEDA PSM is very similar to the *agent-π* model. Thus, the second step of the MDD process is not necessary. In this case, the needed transformation rules are mainly model-to-text, generating directly ANDROMEDA code which can be combined with additional user code. The code template generated by the transformations is shown in Figure 5(a).

(a)

```
public class my_Consumer extends Agent {
    public void init(){
      . . .
     //define agent components
                                                    Behaviour
     Behaviour my_ShopClient= new Behaviour("ShopClient");
                                                    Capability
     Capability my_Service =  new Capability("Service");

     Task my_TaskService =new Task("TaskService");

      . . .
      //add objects into of the agent components
      my_Service.addTaskRun(my_TaskService);
      my_ShopClient.add(my_Service);
      addbehav(my_ShopClient);
   }  // ----------- end init()
                                                    Task
   class my_TaskService extends Task{
      doIt {
        //here the user write the  code
      }  // ----------- end doIt
      . . .
   }  // ----------- end Class Task

   . . .
}  // ----------- end Class Agent
```

(b)

```
public class my_Consumer extends Agent {
   ...                                              Behaviour
  protected void setup(){
     ParallelBehaviour my_ShopClient =
        new ParallelBehaviour( ParallelBehaviour.WHEN_ALL );

     my_ShopClient.addSubBehaviour( new my_Service(this) );
     ...
     addBehaviour(my_ShopClient);
  }  // ----------- end setup()
                                                    Capability
     class my_Service extends OneShotBehaviour {
        public my_Service(Agent a) {
          super(a); }

        public void action() {
          ... //check condition == true
          addBehaviour(new my_TaskService(this) );
        }
     } // ----------- end OneShotBehaviour
    ...                                             Task
     class my_TaskService extends Behaviour {
        public my_TaskService(Agent a) {
           super(a); }

        public void action()  {
            //...this is where the code Task goes !!
        }
     } // ----------- end Behaviour
} // ----------- end class Agent
```

Fig. 5 Code template of ANDROMEDA(a) and JADE-Leap(b) Agents

The transformation from agent models to **JADE-Leap** code must be done employing the two phases previously commented: (i) the first phase translates from the PIM model to JADE-Leap PSM mode, and allows to obtain a correspondence among the abstract concepts of the agent-π model to JADE-Leap concepts; (ii) the second phase allows to translate the JADE-Leap PSM models obtained in the previous phase into executable JADE-Leap code.

In summary, this transformation process converts the `Behaviour` Class of agent-π in a `ParallelBehaviour` Class of JADE-Leap. But also it translates from `Capability` to `OneShotBehaviour` and from `Task` to `Behaviour`. The code template generated by this process is shown in Figure 5(b).

Finally, the code generated automatically by the tool can be edited by the user to complete the *client* agent development with the specific details of the execution platform. Such as, user profiles, i.e., types of products and services preferred by the client that must be manually programmed.

4 Conclusions

This paper has presented the application of the ideas proposed by the MDD for the design of embedded agents into mobile device platforms. Although the use of MDD refers primarily to methodologies for developing object-oriented software, it was verified that the approach can be adopted effectively in the development of agent-oriented software. So, this work has checked the simplification of the agent design, since that some implementation details are hidden to the designer. The presented approach also automates the agent design process because it reduces the human interference in the agent development.

Through the application of MDD, the presented work can design (in a simple and fast way) a virtual organization with embedded agents that run over different platforms (favoring interoperability). The automatic transformations allow generating code templates for different platforms using an unified agent model. Future work of this research will focus on developing new automatic transformations to other embedded agent platforms.

Acknowledgment. This work was partially supported by TIN2009-13839-C03-01 and PROMETEO/2008/051 projects of the Spanish government and CONSOLI-DER-INGENIO 2010 under grant CSD2007-00022.

References

1. Agüero, J., Rebollo, M., Carrascosa, C., Julián, V.: Does android dream with intelligent agents? In: Proceedings of the DCAI 2008, vol. 50, pp. 194–204 (2008)
2. Agüero, J., Rebollo, M., Carrascosa, C., Julián, V.: Agent design using model driven development. In: Proceedings of the PAAMS 2009, vol. 55, pp. 60–69 (2009)
3. Bauer, B.: Uml class diagrams revisited in the context of agent-based systems. In: Proceedings Agent-Oriented Software Engineering, pp. 101–118 (2002)
4. Bergenti, F., Poggi, A.: Leap: A fipa platform for handheld and mobile devices. In: Meyer, J.-J.C., Tambe, M. (eds.) ATAL 2001. LNCS (LNAI), vol. 2333, p. 436. Springer, Heidelberg (2002)
5. Carrascosa, C., Giret, A., Julian, V., Rebollo, M., Argente, E., Botti, V.: Service oriented multi-agent systems: An open architecture. In: Proceedings of the AAMA 2009, pp. 1–2 (2009)
6. Cervenka, R., Trencansky, I.: The Agent Modeling Language – AML. Whitestein Series in Software Agent Technologies and Autonomic Computing (2007) ISBN: 978-3-7643-8395-4
7. Cook, D.J., Das, S.K.: How smart are our environments? an updated look at the state of the art. Pervasive Mob. Comput. 3(2), 53–73 (2007)
8. Hagras, H., Callaghan, V., Colley, M.: Intelligent embedded agents. Information Sciences 171(4), 289–292 (2005)

9. Hahn, C., Madrigal-Mora, C., Fischer, K.: A platform-independent metamodel for multiagent systems. Autonomous Agents and Multi-Agent Systems 18(2), 239–266 (2008)
10. (OMG): Object management group. mda guide version 1.0.1 (June 2003), http://www.omg.org/docs/omg/03-06-01.pdf
11. Pavón, J., Gómez-Sanz, J., Fuentes, R.: Model driven development of multi-agent systems. In: Rensink, A., Warmer, J. (eds.) ECMDA-FA 2006. LNCS, vol. 4066, pp. 284–298. Springer, Heidelberg (2006)
12. Perini, A., Susi, A.: Automating model transformations in agent-oriented modelling. In: Agent-Oriented Software Engineering VI, pp. 167–178 (2006)

Accurate Human Limb Angle Measurement in Telerehabilitation: Sensor Fusion through Kalman, LMS and RLS Adaptive Filtering

Alberto Olivares, J.M. Górriz, J. Ramírez, and Gonzalo Olivares

Abstract. Inertial sensors are widely used in telerehabilitation systems since they permit to monitor the patient's movement and determine the position of their limbs. Limbs angle measurement is carried out through the integration of the angular velocity measured by a rate sensor and the decomposition of the components of static gravity acceleration measured by an accelerometer. Different factors derived from the sensors nature, such as the Angle Random Walk (ARW), and dynamic bias lead to erroneous measurements. Dynamic bias effects can be reduced through the use of adaptive filtering based on sensor fusion concepts. Most existing published works use a Kalman filtering sensor fusion approach. Our aim is to perform a comparative study among different adaptive filters. Several LMS and RLS variations are tested with the purpose of finding the best method leading to a more accurate limb angle measurement. An angle wander compensation sensor fusion approach based on Least Mean Squares (LMS) and Recursive Least Squares (RLS) filters has been developed.

1 Introduction

Due to the overcrowding of rehabilitation centers there is a long waiting list to start most of the treatments after a surgery procedure. When the patient is called out, he is usually scheduled during working hours. Moreover, as a consequence of the overcrowding, patients do not receive a complete treatment and are usually sent home prematurely where they perform the exercises under no supervision.

Alberto Olivares · Gonzalo Olivares
Department of Computer Architecture and Computer Technology,
University of Granada, Spain

J.M. Górriz · J. Ramírez
Department of Signal Processing, Networking and Communications,
University of Granada, Spain

J.C. Augusto et al. (Eds.): ISAmI 2010, AISC 72, pp. 97–104.
springerlink.com

Telerehabilitation improves the efficiency and shortens the duration of the treatment since it allows the patient to remotely carry the exercises without having to renounce to medical supervision [1].

The telerehabilitation system needs instrumentation to monitor the patient, algorithms to process the gathered data, a human-machine interface where the patient can observe in real time the execution of the exercises through a 3D avatar, and a patient-doctor communication platform that allows the latter to supervise the rehabilitation process, check the proper execution of the exercises, update the rehabilitation plan and schedule new sessions and exercises.

The employed instrumentation is composed by inertial sensors, typically accelerometers, gyroscopes and occasionally magnetometers. Within the last decade the use of microelectromechanic (MEMS) sensors has been popularized mainly due to their low size and cost and their reasonable precision. Their market generalization has offered the possibility to develop low cost and low size Inertial Measurement Units (IMUs). However, MEMS devices present a set of negative factors that need to be treated to avoid a loss of precision in the measurements.

One of the main negative effects is the angle wander produced when the gyroscope signal is integrated. This wander is caused by two factors: the first is called Angle Random Walk and consists of random angle wander over time. It is produced when the small peaks which are present in the gyroscope output, due to thermal and electronic noise, are integrated. This effect will vary depending on the quality of the employed sensor. The second factor that contributes to the angle wander is caused by the progressive dynamic variation of the bias in the gyroscope output. The presence of angle wander due to ARW and dynamic bias makes unfeasible the determination of the angle relying solely in the gyroscope.

Therefore, we need a sensor fusion strategy to compensate the negative effects of the angle wander. Sensor fusion is defined as the conjoint use of various sensors to improve the accuracy of the measurements under situations where a sensor is not behaving properly. In our case we have an accelerometer that will help to obtain better measurements.

Accelerometers also present noise in their output, but since we do not have to perform any integration to calculate the angle as it is obtained by the decomposition of the static gravity acceleration, no angle wander is produced. Why do not we always use the accelerometer to measure angles then? The answer depends on the situation in which the limb angles want to be measured. If the patient is moving its limbs slowly, the measured angle will be very accurate as the angle computation relies on static accelerations. By contrast, if the patient is carrying out a rather intense movement, the accelerometer will measure dynamic accelerations leading to totally erroneous measurements.

By using a sensor fusion based approach we will be able to reduce the angle drift taking the accelerometer signal as a position reference that can be weighted through a variable trust parameter according to the degree of dynamic acceleration that we expect the accelerometer to measure.

Such a sensor fusion is usually done using a solution based on Kalman filtering [2]-[4]. Most of the published works tend to use this kind of filters claiming they offer a better response than other solutions. To our knowledge, to this day, there are no published works showing a comparative study among other different adaptive filter sensor fusion approaches.

The goal of our work is to carry out a comparative study between various Least Mean Squares (LMS) and Recursive Least Squares (RLS) filters and the already proposed Kalman filter based approach.

This paper is divided as follows: Section 2 presents the sensor fusion approach based on the Kalman filter and the approach that will allow the use of different adaptive filters. Section 3 presents the obtained results on synthesized signals. Section 4 presents the obtained results on real signals registered by the IMU, and conclusions and future work goals are drawn in Section 5.

2 Sensor Fusion Approaches

As said in the introduction, angles are calculated using the information from acceleration and angular velocity sensed by accelerometers and gyroscopes.

To compute the angle using the accelerometer signal there is no need to perform any integration but just basic trigonometric relations. Since the signal is not integrated the random noise will not cause an angle wander as in the gyroscope case. Given that the angle is computed by sensing the gravity acceleration, we will not be able to obtain a correct measurement when there are dynamic accelerations distorting the sensing of the gravity acceleration. This fact prevents us from using only the accelerometer to measure the angle accurately as short bursts of dynamic accelerations are present even in the less intense activities.

We are forced to use a system that is robust under occasionally presence of dynamic accelerations. If there was no bias drift the gyroscope signal would be very accurate since it does not have high amplitude noise components and is invulnerable to dynamic accelerations. If we manage to remove the bias in the gyroscope signal, we will have a good approximation of the position of the patient's limbs.

In order to remove the bias we can just rely on the only available reference which is the accelerometer angle signal. Since this signal is occasionally erroneous we need a system that is able to decrease the impact of those erroneous observations by filtering them. The proposed tools to perform this task by most authors are adaptive filters and more specifically the Kalman filter [5], [6].

2.1 Kalman Filter Sensor Fusion Approach

The Kalman filter is an ideal solution to perform sensor fusion since it estimates a process based on an observation. The difference between the process estimation and the observation is weighted by a gain which depends on the measurement error covariance and the process error covariance.

The gyroscope signal is filtered based on the acceleration signal observations. We can tell the filter which signal to trust more by modifying the values of the filter parameters. When the measurement error covariance approaches zero, the acceleration signal is trusted more and more, while the predicted measurement is trusted less and less. On the other hand, as the a priori estimate error covariance approaches zero the acceleration signal is trusted less and less, while the predicted measurement using the gyroscope signal is trusted more and more.

This behavior allows us to tune the filter attending to the activity that is being monitored. When monitoring intense activities the accelerometer signal will be given a lower degree of trust to filter dynamic accelerations. On the other hand, if, as in the case of some rehabilitation exercises, activities present low intensity, the accelerometer signal will be trusted over the gyroscope signal.

2.2 LMS and RLS Filters Sensor Fusion Approach

A similar approach is applied when using LMS and RLS adaptive filters.

LMS and RLS filters are widely used to cancel undesired components in signals, such as noise and echo.

We will use the gyroscope signal as the input to be filtered and the accelerometer signal as the desired signal. This configuration will remove the gyroscope's dynamic bias while it will filter the accelerometer's noise, as LMS and RLS will behave as low-pass filters in almost all configurations. The accuracy and convergence time will vary enormously depending on the filter parameters and the input signals nature as is later showed in section 3.

Many variations of LMS and RLS filters have been tested, more specifically, Normalized LMS (N-LMS) [7], Momentum Normalized LMS (MN-LMS) [8], standard RLS, Householder RLS (H-RLS) [9] and QR-decomposition-based RLS (QRD-RLS) [10].

Figure 1 depicts the general diagram of the sensor fusion approach based on adaptive filtering.

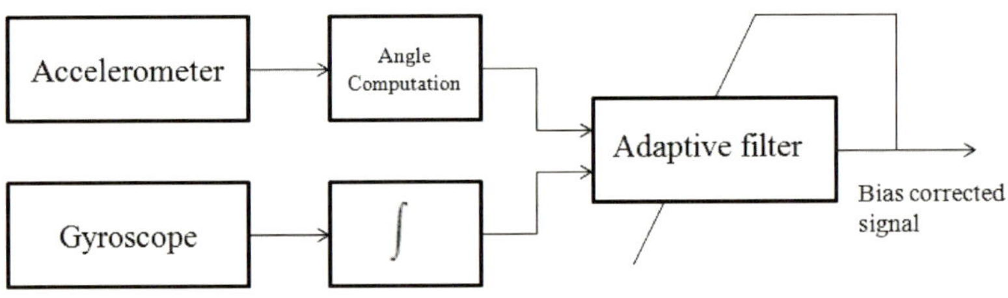

Fig. 1 Diagram of the adaptive filtering sensor fusion approach

3 Theoretical Simulations: Application on Synthesized Signals

In order to perform a comparative study between the adaptive filters it is imperative to synthesize gyroscope and accelerometer angle signals. As the desired unbiased gyroscope signal is unknown, we will synthesize it so the MSE can be performed together with the output of the filter.

Three signals are then synthesized: an accelerometer angle signal, an unbiased gyroscope signal and a biased gyroscope signal.

The unbiased gyroscope signal is synthesized by building a sinusoidal signal having an amplitude of ±180° with a sampling frequency of 50 Hz and a period of 2 seconds. The accelerometer signal is synthesized by adding random noise having a maximum amplitude of 10° to the unbiased gyroscope signal. The biased gyroscope signal is built by adding a linear bias, having the same slope than real measured signals, to the unbiased gyroscope signal. All signals are 8 minutes long. Figure 2 shows the three synthesized signals.

The two synthesized signals are used as inputs of the adaptive filters. The MSE is calculated by averaging the MSE of 50 executions of all filter algorithms. New signals are synthesized for each one of the executions so they will be slightly different as the accelerometer noise has a random nature.

The MSE signals are divided in 3 equal parts for which the average MSE is calculated. These average MSE values will give us information about the filter performance at the beginning, the middle and the end of the time period. Table 1 shows these values.

Based on the simulation results the lowest MSE is obtained for the QRDRLS filter having a filter size of 40 coefficients. Almost all the studied filters outperform Kalman filter in the theoretical simulations. This is due to the poor response offered by the Kalman filter when the dynamic bias value is rather large (period 3).

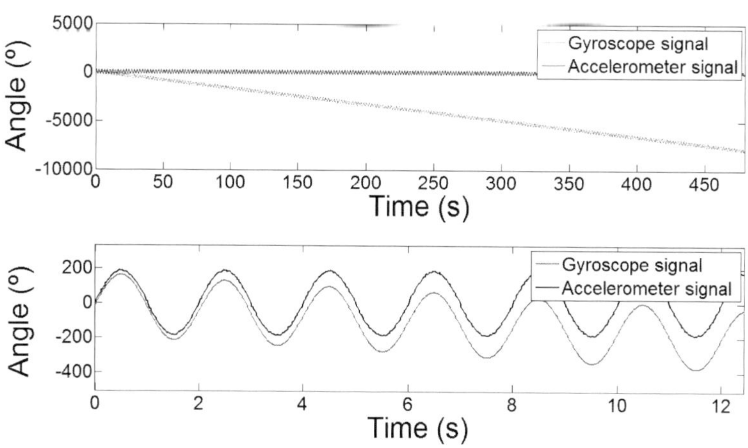

Fig. 2 Synthesized gyroscope and accelerometer angle signals

Table 1 Mean MSE for every of the 3 periods the filter output MSE signal is divided

Filter	Parameters	Mean MSE (dB)		
		Period 1	Period 2	Period 3
Kalman	Sz = 0.03 Var_a = 0.3 Var_g = 0.05	59.64	75.62	84.33
NLMS	Filter size = 600 Step size = 1.5	82.17	74.49	61.48
MNLMS	Filter size = 400 Step size = 1E-6	35.37	26.07	24.9
RLS	Filter size = 10	44.24	23.03	24.48
HRLS	Filter size = 10	47.39	26.42	26.35
QRDRLS	Filter size = 40	36.81	23.33	24.23

Sz: Observation signal noise covariance, Var_a: Process noise covariance related to the acceleration signal, Var_g: Process noise covariance related to the gyroscope signal.

From the table we can observe that, as expected, LMS filters converge slower than RLS filters and, hence, they present a higher MSE during the first period. In the following section the same adaptive filters are applied to real signals gathered by the IMU.

4 Results from Application on Real Signals

After running theoretical simulations to obtain the MSE of each of the adaptive filters we can now apply them to real signals gathered from the accelerometer and gyroscope included inside the IMU. The real signal is obtained from a knee rehabilitation exercise where the IMU is placed on the shank of the patient. The patient moves up and down the shank covering a range of approximately 120º during about 2 minutes. Figure 3 shows the recorded signals for such an exercise.

These two signals are used again as the adaptive filters inputs and the output is calculated for a wide range of values of the different filter parameters. Figure 4 shows the output for the best configurations of each one of the tested filters.

As we can observe in Figure 4, HRLS (filter size=30), RLS (filter size=50) and QRDRLS (filter size=45) offer the best performance as they completely remove the bias and filter quite efficiently the random noise present in the accelerometer signal. Kalman filter does good filtering the noise but does not converge to the proper solution as it presents an increasing positive bias. LMS filters also converge to the desired signal but slower than RLS filters and are also less effective filtering noise.

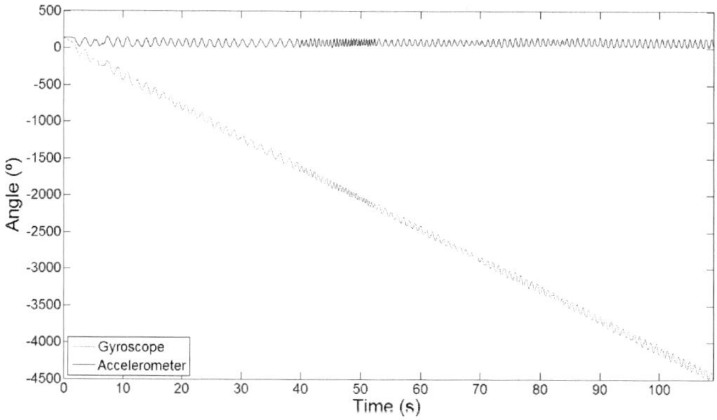

Fig. 3 Accelerometer and gyroscope angle signals gathered during a knee rehabilitation exercise.

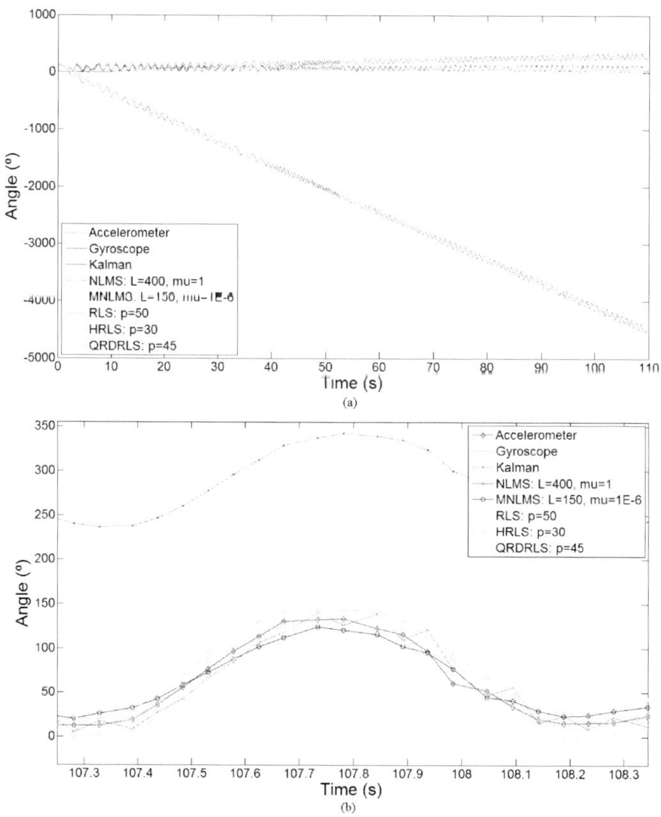

Fig. 4 Output of the adaptive filters for real signals gathered during a knee rehabilitation exercise. Complete signals (a), zoomed signals (b).

5 Conclusions and Future Work

It has been proved that Kalman filter does not converge properly when the dynamic bias effects are large and there is a big difference between the biased and the unbiased gyroscope signals. On the other hand, other adaptive filters such us HRLS, RLS and QRDRLS may be suitable to obtain a more precise solution in systems where the gyroscope presents large dynamic bias.

NLMS is not suitable as its convergence time is too long which causes the first 40 seconds to present totally erroneous measurements. MNLMS does much better in convergence time and might be a good solution to implement in real time systems as its computational cost is many times lower than RLS filters.

Future work will be focused on testing different Kalman filter variations as well as testing the algorithms in a wider range of signals, including those gathered from high intensity activities which will present noise derived from dynamic acceleration in the accelerometer output, and may need from different filters or different filtering parameters.

References

[1] Burdea, G., Popescu, V., Hentz, V., Colbert, K.: Virtual reality-based orthopedic telerehabilitation. IEEE Transactions on Rehabilitation Engineering 8(3), 430–432 (2000)

[2] Luinge, H.J., Veltink, P.H.: Inclination measurement of human movement using a 3-D accelerometer with autocalibration. IEEE Transactions on Neural Systems and Rehabilitation Engineering 12(1), 112–121 (2004)

[3] Lim, Y.P., Brown, I.T.: Kalman Filtering of Inertial Sensors for Evaluation of Orthopaedics Angles. In: ARATA 2008 Conference, Adelaide, Australia (2008)

[4] Yun, X., Lizarraga, M., Bachmann, E.R., McGhee, R.B.: An improved quaternion-based Kalman filter for real-time tracking of rigid body orientation. In: Proceedings IEEE/RSJ International Conference on Intelligent Robots and Systems, vol. 2, pp. 1074–1079 (2003)

[5] Kalman, R.E.: A New Approach to Linear Filtering and Prediction Problems. Transaction of the ASME Journal of Basic Engineering, 35–45 (1960)

[6] Welch, G., Bishop, G.: An introduction to the Kalman filter. Notes of ACM SIGGRAPH tutorial on the Kalman filter (2001)

[7] Widrow, B., Glover, J.R., Mccool, J.M., Kaunitz, J., Williams, C.S., Hean, R.H., Zeidler, J.R., Dong, E., Goodlin, R.C.: Adaptive noise cancelling: Principles and applications. Proc. IEEE 63(12), 1692–1716 (1975)

[8] Douglas, S.C., Meng, T.H.Y.: Normalized data nonlinearities for LMS adaptation. IEEE Trans. Signal Process 42(6), 1352–1354 (1994)

[9] Athanasios, A.: Rontogiannis and Sergios Theodoridis On inverse factorization adaptive least-squares algorithms. Signal Processing 52(1), 35–47 (1996)

[10] Alexander, T.S., Ghirnikar, A.L.: A Method for Recursive Least Squares Filtering Based Upon an Inverse QR Decomposition. IEEE Transactions on Signal Processing 41(1), 20–30 (1993)

An Agent-Based Framework for Context-Aware Digital Signage

Ichiro Satoh

Abstract. This paper presents a context-aware infrastructure for building and managing mobile agent-based content on digital signage. These are digital representations of people or physical entities and they can define and provide multimedia content for advertising or user-assistant services for users. The infrastructure spatially binds the positions of people or objects with the locations of their virtual counterparts and, when they move or are transported in the physical world, it automatically deploys their agents at computers near their current positions to provide ambient media.

1 Introduction

Digital signage has been expected to play an important role in advertising, navigation, assistance, or entertainment in public and private environments, such as that at public museums, retail stores, and corporate buildings. Advertising using digital signage should be a form of out-of-home advertising in which content and messages are displayed on digital signs with a common goal of delivering targeted messages specific consumers in specific locations at specific times. For example, most visitors to public museums want annotations on the exhibits in front of them, because they lack sufficient breath of knowledge about the exhibits. Their knowledge and experiences are varied so that they tend to be become puzzled (or bored) if the annotations provided to them are beyond (or beneath) their knowledge or interest. Context-awareness should provide users with services according to the users' context. This paper presents a framework for building and managing context-aware advertising on digital signage in public places, in-transit, or waiting. We have four key ideas behind this.

Ichiro Satoh
National Institute of Informatics
2-1-2 Hitotsubashi, Chiyoda-ku, Tokyo 101-8430, Japan
e-mail: ichiro@nii.ac.jp

J.C. Augusto et al. (Eds.): ISAmI 2010, AISC 72, pp. 105–112.
springerlink.com

- Content, which is played on digital signage, can be anything, including text, images, animations, video, audio, and interactivity. The framework should provide a variety of multimedia content, including interactive content material to consumers, from digital signage.
- The framework should monitor contextual information in the real world by means of sensing systems and select and adapt content according to changes in the real world. to enhance customer experience.
- The framework needs to provide massive numbers of users with context-aware services at massive numbers of ubiquitous and mobile computing devices in personalized form.
- It should provide context-aware content with the ability to monitor and respond to users' behaviors, because advertising or user-assistant services provided as context-aware services should adapt to the users' profiles.

This framework introduces the notion of mobile agents, where mobile agents are autonomous programs that can travel from computer to computer under their own control as virtual counterpart objects of people or physical entities. No physical objects, including legacy appliances, may specify their attributes or interact only with one another, because of limited resources. An application cannot directly interact with people or physical objects but with their virtual counterpart agents. Their counterpart agents interact with one another on behalf physical objects.

The framework also introduces the notion of context-awareness as a bridge a gap between the physical and virtual worlds, which consists of virtual-counterpart agents. It spatially binds the positions of people or objects with the locations of their virtual counterparts and, when they move or are carried in the physical world, it automatically deploys their agents at computers near to their current positions to provide services. As context-aware services should in widespread throughout buildings and cities, they cannot be managed by using traditional approaches, such as those that are centralized and top-down. However, most existing attempts at context-aware services have been implemented in an ad-hoc and centralized manner.

There have been several commercial projects that have focused on providing context-aware content on digital signage, but they have been constructed based in an ad-hoc manner. However, several researchers have explored context-aware services independently of the literature on digital signage. Cambridge University's Sentient Computing project (HHSWW99) provides a platform for location-aware applications using infrared-based or ultrasonic-based locating systems in a building. Microsoft's EasyLiving project (BMKKS00) enabled services running on different computers to be combined dynamically according to contextural changes in the real world. We discuss differences between the framework presented in this paper and our previous frameworks. We constructed a location model for ubiquitous computing environments. The model represented spatial relationships between physical entities (and places) as containment relationships between their programmable counterpart objects and deployed counterpart objects at computers according to the positions of their target objects or places (Satoh05). This was a general-purpose location-model for context-aware services, but was not an infrastructure for deploying and operating such services. We presented an outline of mobile agent-based services in public museums in our early version of this paper (Satoh09), whereas this paper addresses agent-based advertising on digital signage for shopping.

2 Basic Approach

This framework builds and manages context-aware multimedia content on digital signage. It enables such content to be implemented as mobile agents so that they can be dynamically deployed at computers close to users according to context in the real world, e.g., the locations of users and physical objects detected by using locating systems.

- They need to be managed in a non-centralized manner. Mobile agents can be managed without any centralized servers to support large-scale context-aware systems.
- All mobile agents are programmable entities with data. Therefore, all mobile agent-based services can define what programs play their visual/audio content and interact with their users. Therefore, the framework itself is independent of application-specific tasks and provides multiple kinds of multimedia content, because such tasks are performed within mobile agents.
- After arriving at their destinations, mobile agents can continue their work without losing the results of working, e.g., the content of instance variables in the agents' programs, at the source computers. Therefore, users can continue to watch or listen to content from computers close to their current positions, even when they move from location to location.
- The framework deploys and executes mobile agents bound to physical entities or people at computers near the positions of the users instead of at remote servers. As a result, mobile agent-based content can directly interact with users, where RPC-based approaches, which other existing approaches are often based on, must have network latency between computers and remote servers.

3 Design and Implementation

The framework consists of three parts: (1) mobile agents, (3) agent runtime systems, and (3) location information servers, called LISs (Fig. 1). The first offers application-specific content, which is attached to physical entities and places, as collections of mobile agents. The second runs on digital signage and is responsible for executing and migrating mobile agents. The third provides a layer of indirection between the underlying locating-sensing systems and mobile agents. Each LIS manages more than one sensor and provides the agents with up-to-date information on the state of the real world, such as the locations of people, places, and things, and the destinations that the agents should migrate themselves to.

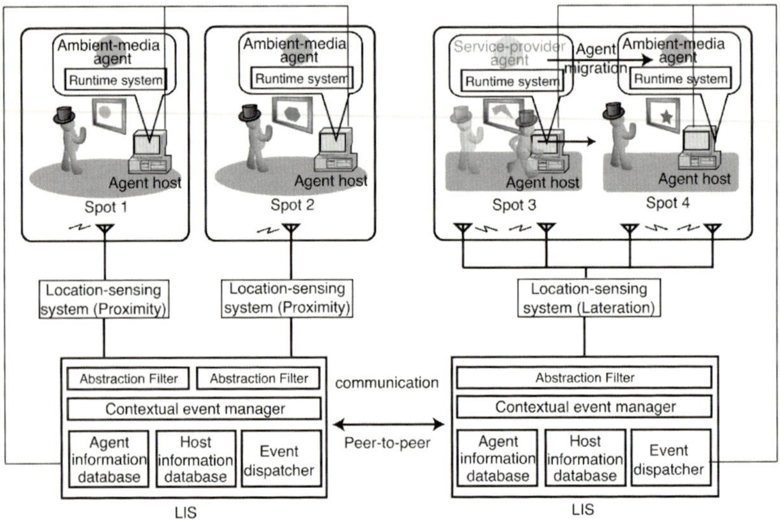

Fig. 1 Architecture

3.1 Location Information Server

Each LIS can run on a stationary or mobile computer. It manages multiple sensors that detect the presence of tags and maintains up-to-date information on the identities of tags that are within the zone of coverage of its sensors. This is achieved by polling the sensors or receiving the events issued by the sensors themselves. An LIS does not require any knowledge of other LISs. To conceal the differences between the underlying locating systems, each LIS maps low-level positional information from each of the locating systems into information in a symbolic model of location. An LIS represents an entity's location, called *spot*. e.g., spaces of a few feet , which distinguishes one or more portions of a room or building. When an LIS detects a new tag in a spot, it multicasts a query that contains the identity of the new tag and its own network address to all the agent runtime systems in its current sub-network to discover agents. When the absence of a tag is detected in a spot, each LIS multicasts a message with the identifier of the tag and the identifier of the spot to all runtime systems in its current sub-network.

3.2 Agent Runtime System

Each runtime system migrates agents to other runtime systems running on different computers through a TCP channel using mobile-agent technology. It is built on the Java virtual machine (Java VM) version 1.5 or later, which conceals differences between the platform architectures of the source and destination computers. It governs all the agents inside it and maintains the life-cycle state of each agent. When the life-cycle state of an agent changes, e.g., when it is created, terminates, or migrates to another runtime system, its current runtime system issues specific events to the agent.

When an agent is transferred over the network, not only its code but also its state is transformed into a bitstream by using Java's object serialization package and then the bit stream is transferred to the destination. Since the package does not permit the stack frames of threads to be captured, when an agent is deployed at another computer, its runtime system propagates certain events to to instruct it to stop its active threads. Arriving agents may explicitly have to acquire various resources, e.g., video and sound, or release previously acquired resources. The framework only maintains per-user profile information within those agents that are bound to the user.

3.3 Mobile Agent-Based Multimedia Content

Each agent is attached to at most one visitor and maintains his/her preference information and programs that provide annotation and navigation to him/her. Each agent keeps the identifier of the tag attached to its visitor. Each agent is dynamically assembled from the *content part* and *user-preference part* to support user/location-dependent content.

The content part is responsible for selecting and playing annotations according to the current spot and route in addition to the information stored in the user-preference part and it plays the content in the personalized form of its user. It is defined as a set of content-selection functions and programs for playing the selected content.

The function maps more than one argument, e.g., the current spot, the user's selected route, and the number of times he/she has visited the spot into a URL referring to the annotative content. The content can be stored in the agent, the current runtime system, or external http servers. That is, each agent can carry a set of its content, play the selected content at its destinations, directly play the content stored at its destinations, or download and play the content stored in Web servers on the Internet. The current implementation can divide this part into three sub-parts: opening, annotation, and closing, which are played in turn.

Annotation content is various, e.g., text, image, video, and sound. The annotation part defines programs for playing this content. The current implementation supports (rich) text data, html, image data, e.g, JPEG and GIF, video data, e.g., animation GIF and MPEG, and sound data, e.g., WAV and MP3. The format for content is specified in an MIME-based attribute description. Since the annotation part is defined as Java-based general-purpose programs, we can easily define interactions between visitors and agents.

The user-preference part is responsible for maintaining information about a visitor. In fact, it is almost impossible to accurately infer what a visitor knows or is interested in from data that are measured by sensing systems. Instead, the current implementation assumes that administrators will explicitly ask visitors about their knowledge and interests and manually input the information into this part. Nevertheless, it is still possible to make an educated guess with some probability as to what a visitor may be interested in, if we know which spots he/she has visited, how many times he/she visited them, and how long he/she spend in these visits. Each agent has a mechanism to automatically record the identifiers, the number of visits to, and the length of stays at spots by visitors.

3.4 Remarks

The current implementation supports a commercial tracking system, called a Spider active RFID tag system, which is a typical example of proximity-based tracking. We measured the cost of migrating a null agent (a 5-KB agent, zip-compressed) and an annotation agent (1.2-MB agent, zip-compressed) from a source runtime system to a destination runtime system that was recommended by the LISs. The latency of discovering and instructing an agent attached to a tag after the CDS had detected the presence of the tag was 420 ms and the respective cost of migrating the null and annotation agent between two runtime systems over a TCP connection was 38 ms and 480 ms.[1]

4 Application

We experimented on and evaluated mobile agent-based active media for appliances, e.g., electric lights. This is unique to other existing active media because it does not support advertising of its target appliances but assists users in controlling and disposing of the appliances. We attached an RFID tag to an electric light and provide a mobile agent as an active medium for the light. The media is attached to its target item and were deployed at computers close to the current position of the item. The current implementation assumed that an agent the managing active media for its target appliance created when the appliance was shipped from the factory. Since the agent defines programs to display three kinds of active media content inside it, it selects them according to their spaces. It supports the lifecycle of the item from shipment, showcasing, installation, usage, and disposal.

- **In warehouse:** While the light is in a warehouse, its agent is deployed at a computer in the warehouse. It notifies to the warehouse's server of its specifications, e.g., its product number, its serial number, its date of manufacture, the size and weight,
- **In store:** While the light is being showcased in a store, its agent is deployed at a computer close to its target object so that it can display advertising media to assist sales to customers who are visiting the store. Figures 2 a) and b) are two images that are maintained in the agent to display the price, product number, and manufacture's name on its current computer.
- **In house:** When the light is bought and transported to its purchasers' house, its agent migrates to a computer in the house and explains how it is installed. Figure 2 c) has the active media to guide assembly. The agent also explains how

[1] This evaluation was assessed with three computers (Intel Core 2 Duo 1.8 GHz with Mac OS X 10.5 and JDK 1.5) connected via a Fast Ethernet.

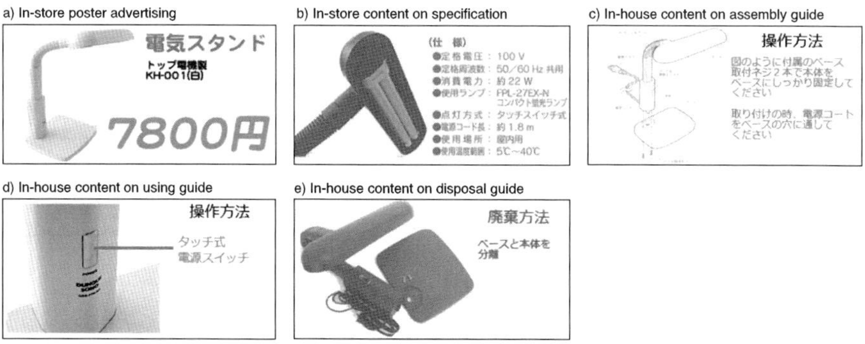

Fig. 2 Digital signage to support appliance

it is used as shown in Figure 2 d). When it is disposed of, the agent displays its active media to guide disposal. Figure 2 e) has an image that illustrates how the appliance is disposed.

In the setting of a house, we can define agents that control appliances, which may not have any network interfaces. In both of the approaches we describe here, the lights are controlled by switching their power sources on or off through a commercial protocol, called X10.

The first can autonomously turn room lights on whenever a user with a tagged user is sufficiently close to them. The agent attached to the light can also work as our X10-based server's client and it runs on the stationary runtime system in the room. When a tagged user approaches a light, an LIS in the room detects the presence of his/her tag in the cell that contains the light. The LIS then moves the agent that is bound to his/her tag to the runtime system on which the light's agent is running. The user's agent then requests the lights' agent to turn the light on through inter-agent communication.

The second approach allows us to use a PDA to remotely control nearby lights. Place-bound controller agents, which can communicate with X10-base servers to switch lights on or off, are attached to places with room lights in this system. Each user has a tagged PDA, which supports the runtime system. When a user with a PDA is visiting a cell that contains a light, the framework moves a controller agent to the runtime system of the visiting PDA. The agent, now running on the PDA, displays a graphical user interface to control the light. When the user leaves that location, the agent automatically closes its user interface and returns to its home runtime system.

5 Conclusion

We designed and implemented a context-aware infrastructure for building and managing mobile agent-based content displayed on digital signage, where mobile agents are autonomous programs that can travel from computer to computer under their own control as virtual counterpart objects of people or physical entities. It provides users and physical entities with mobile agent-based content to support and annotate them. Using location-tracking systems, it can migrate content to stationary or mobile computers near the locations of the users and physical entities to which the agents are attached. That is, it allows a mobile user to access its personalized services in an active computing environment and provides user/location-dependent active media to a user's portable computer or stationary computer. It is managed in a decentralized manner.

Acknowledgements. This research is supported by Promotion program for Reducing global Environmental loaD through ICT innovation (PREDICT), Ministry of Internal Affairs and Communications of Japan.

References

[BMKKS00] Brumitt, B.L., Meyers, B., Krumm, J., Kern, A., Shafer, S.: EasyLiving: Technologies for Intelligent Environments. In: Proceedings of International Symposium on Handheld and Ubiquitous Computing, pp. 12–27 (2000)

[HA03] Harrison, J.V., Andrusiewicz, A.: Enhancing digital advertising using dynamically configurable multimedia. In: 2003 International Conference on Multimedia and Expo (ICME 2003), vol. 1, pp. 717–720 (2003)

[HHSWW99] Harter, A., Hopper, A., Steggeles, P., Ward, A., Webster, P.: The Anatomy of a Context-Aware Application. In: Proceedings of Conference on Mobile Computing and Networking (MOBICOM 1999), August 1999, pp. 59–68. ACM Press, New York (1999)

[Satoh05] Satoh, I.: A Location Model for Pervasive Computing Environments. In: Proceedings of IEEE 3rd International Conference on Pervasive Computing and Communications (PerCom 2005), March 2005, pp. 215–224. IEEE Computer Society, Los Alamitos (2005)

[Satoh09] Satoh, I.: A Context-aware Service Framework for Large-Scale Ambient Computing Environments. In: Proceedings of ACM International Conference on Pervasive Services (ICPS 2009), July 2009, pp. 199–208. ACM Press, New York (2009)

What Context Matters? Towards Multidimensional Context Awareness

Víctor Penela, Carlos Ruiz, and José Manuel Gómez-Pérez

Abstract. Context management has been a key technique in Ambient Intelligence applications development, although its use has not attain such success in other environments, such as those evolved from the Web 2.0 paradigm and the ubiquity of the mobile devices, two cases where social relationship information can be used for improving context providing and consuming. This paper proposes a context relevancy vision and a set of context dimensions in order to improve context information processing and distribution.

1 Introduction

Context most accepted definition comes from Dey and Abowd[3] describing it as "Context is any information that can be used to characterize the situation of an entity. An entity is a person, place or object that is considered relevant to the interaction between a user and an application, including the user and the application themselves.". That definition provides with a first glimpse on one of the biggest issues that context-aware applications face: any item of information could, at some point and under some conditions, be seen as contextually relevant or not. Although logical from a theoretical point of view, in practice it becomes totally unscalable for context-aware applications. The amount of information available nowadays is growing so fast that it makes it impossible to leverage and consume all that data, even from a design point of view. Furthermore, the relevancy of each context source or even of each context information item is not static, particularly in mobile environments or when building platforms where applications from different domains could act as context consumers or providers, so a model too rigid or too domain dependent will only solve particular problems, but not provide with a reusable view of the whole issue.

Víctor Penela · Carlos Ruiz · José Manuel Gómez-Pérez
iSOCO, Pedro de Valdivia 10, Madrid, Spain
e-mail: {vpenela,cruiz,jmgomez}@isoco.com

J.C. Augusto et al. (Eds.): ISAmI 2010, AISC 72, pp. 113–120.
springerlink.com

The classical standpoint to solve this issue involves filtering the context information provided by physical variables, such as the distance between source and consumer (the context data location), and the longevity of the the context information item (the context data freshness), both already implemented by the Context Toolkit[13]. Other context dimensions should be taken into account, such as social relations and the role of the consumer. As an example, the user of a social application will find knowing that a friend is busy at a particular time quite useful when planning to go to a movie. From a role point of view, two people that share a common work background even though not related to each other or to the same company, e.g. two financial analyst working in different companies and different countries, will probably share to some degree the same interest in a particular set of information sources, the valuation of different trade markets.

The paper is as follows: Section 2 overviews the current state of art. Section 3 stresses the importance of a key element such as relevancy to define what is context. Section 4 extends the concept of context dimension and the mechanisms for its calculation. Section 5 provides with particularized examples of how this approach can improve user experience. Finally, section 6 presents the conclusions and future lines of research.

2 State-of-Art in Context Relevancy

There have been strong research in the last years on context relevancy, focused mainly in two areas, how to reduce data volume and how to share it with others.

Context processing aims to reduce the amount of information taken into account. A hierachy of contexts is introduced in [1], seeing them as subsets of all the available environmental inputs. However, which inputs are selected for each subset is done in the design phase, a process that is far from optimal in a fast changing environment. Context Summarization[12] reduces the size of the context dataset, by converting raw data into smaller processed sets via methods such as aggregation or pattern identification. However this process has to be defined in a case by case scenario.

On the other hand, although it can lead to an increase in the final amount providing with a social background for context sharing[5] is an absolutely needed capability in context-aware applications. ContextPhone[11] lets users share context information between them, however it strongly relies in location proximity for that sharing to be considered relevant. Eliminating the location restriction, Context Watcher[7] lets users provide context information to their "buddies", but restrict the approach to informating strongly related to mobile devices usage, such as location, agenda and photo sharing and relations need to be established in their own and independent social network.

In many cases the relevancy of an environmental input is decided at design time, without any dynamic adaptation at execution time. The same variable could or not be relevant in a particular situation depending on the rest of the available inputs, so a static approach could result in a poor performance of the context retrieval process.

As well, most of the approaches apply only to prearranged scenarios where all the context sources are well known and can be set up as needed, being quite impossible to extend them to open world problems. Because of that, applications built that way are hardly usable out of the original environment for which they were designed. Although it is quite impossible to build a system that perfectly fits any given domain, the mechanisms proposed should go beyond a mere ontological design and try to tackle the big picture issue.

3 Context Relevancy

Dourish[4] defines context as an interactional problem between objects or activies, and thus some data can not simply be or not be context, but contextually *relevant* or not. The scope of that relevancy must always be defined dynamically.

Evaluating context relevancy is a keystone for improving user experience in context-aware applications. Most of the previous efforts translate relevancy to what can be seen as *physical relevancy*, i.e. sources physically close to the consumer. These approaches do not take into account social information, although it is one of the most important sources of added value information for the final consumer, specially in applications and services built around the Web 2.0 paradigm[10]. A parallelism between the information exchange that a network of sensors in a close range to the consumer and the social graph that a consumer and its connections made (seeing these connections as possible context information sources) will be used to define a possible mechanism to extend the original model. From the same perspective, two consumers sharing a particular set of characteristics, thus being seen as part of the same group or cluster, although not directly related to each other, would probably share the same interest in particular types of context information.

Introducing these two concepts could enhance the information management processes involved in context aware applications, allowing to extend the range of fields where they could be applied. In particular, this approach will not only need to include social and role concepts in its model as relevant aspects, but it must be designed as a system able to exploit such concepts in order to improve the final consumption of context-aware applications.

Most of the current approaches on Ambient Intelligence limit themselves to what could be called Ambient Context, doing no filtering at all or filtering only by the position of the source ("leverage only sources that are in a radius of 2 kilometers") or the timestamp when the item of context information was produced ("ignore any information older than 20 minutes"). The solution is transcending that classical point of view and expanding the vision of context to an *Extended Context*, where others types of information, such as social or role-based, are also taken into account.

The main property an entity has to possess in order to be considered part of the context is its relevance to the final context consumer. This criterion is at the end an objective way of implicitly weighting how a piece of information is important for a particular consumer.

$$C_i = \phi(S_J) \quad (1)$$

As shown in 1, context of a user i seen as C_i is a function ϕ that defines how relevant a context source S_j is, being j an individual context source and J the whole set of existent context sources. This approach assumes that the items in J are known, although that proposition is made only for simplicity, and the results can be extended to a more realistic situation where context sources are not completely known from the start. Optimizing this process, and thus the calculus of ϕ, will at the end improve the quality of the gathered context and enrich the context-aware user experience.

4 Context Dimensions

Previous efforts have tried to decompose context in different dimensions in order to be able to define its key concepts in a simpler way. Lenat[8], under the premise of seeing the context as "a region in some n-dimensional space", defines a 12-dimensional space using the Cyc knowledge base that provides with a superstructure where different contexts can be specified. However Lenat uses this approach for classifying a universal and global context, while in this paper the goal relevancy is defined in a item per item basis.

Kappel[6]proposes a context model that divides information in two layers: *physical context model* and *logical context model*. While physical properties are defined at a very low level of abstraction, logical profiles are built by increasing the abstraction level. Nevertheless, this approach is designed only at a modeling level, not developing the necessary methods to define the boundaries of a particular context.

As well, Buchholz[2] proposes a set of parameters to define *Quality of Context*, that includes *Precision*, *Probability of Correctness*, *Trust-worthiness*, *Resolution* and *Up-to-dateness*. However, although most of this parameters are not domain dependet, they are strictly related to physical qualities of the source or the measurement by itself, but not extended to higher relations between source and consumer.

Several dimensions are proposed next, although the final goal is to provide with a big picture vision by defining a common framework where other dimensions can be evaluated allowing for the optimization of context management mechanisms. The main characteristics a magnitude needs to cover to be considered a context dimension are:

1. Domain independence. A dimension should not be restricted to an specific domain, i.e. only being able to define relevancy for context sources that provide information in a particular domain.
2. Dynamic adaptability. A dimension should define a set of parameters that are not static. Defining static parameters will be another way of only applying relevancy criteria during the design phase, which is, in most of the cases, not useful for a dynamic context-aware application.

Ambient Dimension

Probably the most usual way of determining the relevancy of an input is by analyzing the physical properties related to an specific context source. Different magnitudes could be used for filtering inputs from a physical point of view, being the most important position and time.

Only those inputs with an absolute or relative specific position are considered relevant. That way a subset of sources in a specific place, e.g., Madrid or 40 26N 3 42W, or simply put, close to the context consumer, e.g., sources in a radius of 5 kilometers could be filtered in. As well, context relevancy is strongly coupled with time, mainly by the amount of time since it was gathered, e.g. only data retrieved less than 5 minutes ago will be used.

Social Dimension

Context could be extended by defining a user's context as the original environmental context and the individual contexts of the user's social graph members. From this point of view, context relevancy must be also seen as a way of limiting what information of other user's context is relevant to the user per se. Optimizing the selection of this subset of relevant inputs, and, moreover, the selection of the subgraph of the social graph of that consumer that is relevant in a particular situation (thus stressing the dynamic adaptability of the social dimension) are not trivial tasks. The different methodologies and approaches to these tasks will not be evaluated in this study as they have been researched more exhaustively in previous works [15][9].

Using the concepts of social graph and the notation from [15] a relationship between two nodes n_i and n_j in a X_r relation is defined as x_{ijr} and the distance between, seen as the number of edges of the graph it has to walk to get from one node to the other, as l_{ijr}. Both R (the type of network in which this connection is specified), X (the particular kind of relation in the R domain) and L (the distance between two nodes in a R domain) can be used to add relevancy to a context item produced by a context source n_i affecting a context consumer n_j.

Taking these three parameters into account, a social subgraph can be defined as a subset of all the existent social relationship a particular context consumer possess.

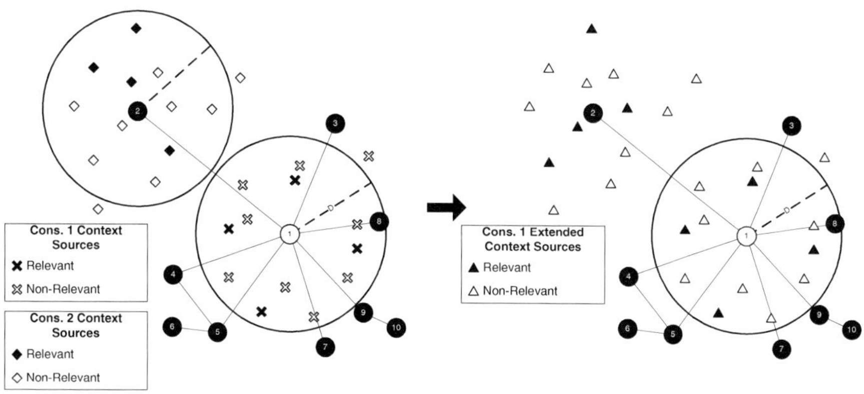

Fig. 1 Extendede Social Context from one connection in the social graph

How this subset is selected is not a trivial question, although its solution is not a matter for this study, so we will suppose that this specific graph exists and is known.

Again, particularizing 1, ϕ function can be seen as where a variable β_{ijs}, that defines how relevant for a context consumer i a context source j in the frame of a specific social subgraph s.

$$C_i = \sum_{i=1}^{J} \beta_{ijs} S_j \tag{2}$$

Role Dimension

Playing a role[14] will modify the relevance of the context inputs the user is exposed to, and that modification will affect in the same way any user playing that role.

$$C_i = \sum_{i=1}^{J} \delta_{ijr} S_j \tag{3}$$

The relevancy of a context source S_j for users playing a role r (from a family R of roles) is defined by γ_{jr}. That definition takes into account that a role will introduce the same modifying factor to any context consumer playing it. Adding that effect to the initial relevancy a context source had for the consumer, a consumer C_i playing a role r will have a context noted as C_{ir}, defined by 3, where δ extends γ.

5 Application Scenarios

Current real world applications raise new challenges in context-aware systems that are not covered in classic Ambient Intelligence approaches. The following scenarios are presented trying to use real world situations to show the improvement margin that can be achieved by maturing these techniques and enriching the context management process with social and behavioral information as context relevancy rating criterion.

All the scenarios are designed deliberately small and parallel to show the huge difference that a relevancy weight mechanism can provide for context-aware applications even when the changes between cases are so small. The specific information or services consumed are not relevant. The main goal in the three of them is to provide the user with the more relevant context information for each of them, taking into account the differences between users, which, although subtle, can be totally game-changing in choosing which pieces of information and sources are selected.

1. User A is a tourist in *Madrid*, visiting the city for the first time.
2. User B is a tourist in *Montreal*, visiting the city for the first time.

From a classical point of view, users A and B are in different locations that define different environment inputs. The context of User A will be defined by those sources of information located in Madrid (or in a radius R of the consumer), while that of User B will be built with information provided by sources close to his location, Montreal.

1. User A is a tourist in Madrid with *no contacts in the city*.
2. User B is a tourist in Madrid with *several friends in the city*.

By applying only ambient filtering both users will have exactly the same context, since both their location and the environment inputs surrounding them are exactly the same. However, from a social perspective, the user experiences of both of them could be quite different, which should be reflected in the service received. While user A has no contacts in the city, user B has social connection that can provide him with extended environment information (e.g., personal opinions about places) or with social information (e.g., their schedules in order to arrange a meeting together).

1. User A is a *tourist* in Madrid. He is doing a tour of the city by bus.
2. User B is a *bus driver* in Madrid. He is the tour of the city bus driver.

As explained in the previous point, from a classical point of view, both users will receive exactly the same context information. However, it is clear that the pieces of information that each of them would feel more relevant are quite different: while user A would probably be more interested in sources of information from the city historic places and sightseeing spots, user B would find traffic information more relevant.

6 Conclusions

In this paper a methodology for evolving context providing by taking into account the concept of context relevancy. Context relevancy has been implicitly used since the context-aware computing origins, but an explicit model has never been proposed. With this model both context filtering and context source prioritizing can be easily defined.

Furthermore, context relevancy is the start point for evolving the classical Ambient Intelligence context concept into what has been called Extended Context, that takes into account information provided by context sources that are not explicitly related to the user from a environmental point of view.

A model has been provided to evaluate context information relevancy to context consumer taking into account social connections and roles. This model will be built into specific applications for practical testing purposes. That way a more explicit model for the calculus of the Extended Context will be designed, allowing for a higher level of optimization in the context management process.

As well, two dimensions have been proposed for extending the Classical Context model into a Extended Context. Both the social and the role dimensions result from previous work on the field and the needs that were felt as lacking from the classical conception of Ambient Intelligence. However, other dimensions could be extracted from different scenarios and added to the general model.

Acknowledgements. This research work is being supported by the CENIT Spanish National Research Program, as part of the INGENIO2010 Spanish National Fund.

References

1. Biegel, G., Cahill, V.: A framework for developing mobile, context-aware applications. In: IEEE International Conference on Pervasive Computing (PerCom), p. 361. IEEE Computer Society Press, Los Alamitos (2004)
2. Buchholz, T., Küpper, A., Schiffers, M.: Quality of context: What it is and why we need it. In: Proceedings of the Workshop of the HP OpenView University Association (2003)
3. Dey, A., Abowd, G.: Towards a better understanding of context and context-awareness. In: CHI 2000 Workshop on the What, Who, Where, When, and How of Context-Awareness, pp. 304–307 (2000)
4. Dourish, P.: What we talk about when we talk about context. Personal and ubiquitous computing 8(1), 19–30 (2004)
5. Hofer, T., Schwinger, W., Pichler, M., Leonhartsberger, G., Altmann, J., Retschitzegger, W.: Context-awareness on mobile devices-the hydrogen approach. In: Proceedings of the 36th Annual Hawaii International Conference on System Sciences, pp. 292–302 (2002)
6. Kappel, G., Proll, B., Retschitzegger, W., Schwinger, W.: Customisation for ubiquitous web applications: a comparison of approaches. International Journal of Web Engineering and Technology 1(1), 79–111 (2003)
7. Koolwaaij, J., Tarlano, A., Luther, M., Nurmi, P., Mrohs, B., Battestini, A., Vaidya, R.: Context watcher: Sharing context information in everyday life. In: Proceedings of the IASTED conference on Web Technologies, Applications and Services, pp. 12–21 (2006)
8. Lenat, D.: The dimensions of context-space. Tech. rep., Cycorp (1998)
9. Newman, M.: Detecting community structure in networks. The European Physical Journal B-Condensed Matter and Complex Systems 38(2), 321–330 (2004)
10. O'Reilly, T.: What is web 2.0: Design patterns and business models for the next generation of software (2005), `http://oreilly.com/web2/archive/what-is-web-20.html`
11. Raento, M., Oulasvirta, A., Petit, R., Toivonen, H.: Contextphone: A prototyping platform for context-aware mobile applications. IEEE Pervasive Computing 4(2), 51–59 (2005)
12. Rasheed, F., Lee, Y., Lee, S.: Towards using data aggregation techniques in ubiquitous computing environments. In: IEEE international Conference on Pervasive Computing, pp. 369–372 (2006)
13. Salber, D., Dey, A., Abowd, G.: The context toolkit: Aiding the development of context-enabled applications. In: Proceedings of the SIGCHI conference on Human factors in computing systems: the CHI is the limit, p. 441 (1999)
14. Steimann, F.: On the representation of roles in object-oriented and conceptual modelling. Data & Knowledge Engineering 35(1), 83–106 (2000)
15. Wasserman, S., Faust, K.: Social network analysis: Methods and applications. Cambridge University Press, Cambridge (1994)

A Framework for Modeling and Analysis of Ambient Agent Systems: Application to an Emergency Case

Tibor Bosse and Alexei Sharpanskykh

Abstract. It is recognized in Ambient Intelligence that ambient devices should be modeled as intelligent autonomous components rather than passive information sources. The agent paradigm suits well for representation of both intelligent ambient devices and humans. However, modeling complex cognitive dynamics from which intelligent behavior of agents emerges is not trivial. In this paper a formal framework is proposed for declarative modeling of cognitive processes and behavior of agents in an ambient intelligence system. Models obtained by the framework are executable and can be used directly for simulation and analysis. The framework provides rich analysis possibilities of qualitative and quantitative properties, specified at local (agent) and global (system) levels. An application of the framework in the context of a subway evacuation case is described in the paper.

1 Introduction

It is often argued in Ambient Intelligence that ambient devices should be modeled as intelligent autonomous components rather than passive information sources (cf [10]). The agent paradigm suits well for representation of intelligent ambient devices [5]. Agents are autonomous entities able to make decisions and interact with the environment by communication, observation and actions. Externally observable behavior of intelligent agents emerges from their complex cognitive dynamics. The cognitive dynamics comprises cognitive processes which, as widely recognized in the literature [6], influence intelligent agents' decision making: e.g., belief revision, learning, dynamics of trust, generation and development of emotions.

In general, modeling of ambient agent systems is a challenging task. Models of such systems should account for complexity of human behavior, heterogeneity and autonomous behavior of technical systems, and interaction among all system

Tibor Bosse · Alexei Sharpanskykh
VU University Amsterdam, De Boelelaan 1081a Amsterdam, the Netherlands
e-mail: {tbosse,sharp}@cs.vu.nl

J.C. Augusto et al. (Eds.): ISAmI 2010, AISC 72, pp. 121–129.

actors. Both quantitative and qualitative system properties are required to be modeled. Furthermore, relations between diverse processes and states of different aggregation levels of a system should be defined explicitly. In this paper we propose a formal framework, which addresses all these modeling requirements and provides means for performing various simulation experiments with ambient intelligence systems. In contrast to other existing approaches and tools for modeling and analysis of ambient intelligence systems (cf [8], [7], [9]), the framework allows declarative modeling of detailed cognitive processes of agents including belief revision, trust and emotion development, and decision making. This is also a novel contribution of this paper. Furthermore, the framework enables modeling externally observable behavior of agents such as communication between agents and physical actions of agents in the environment (e.g., movement). The structure and dynamics of the environment can be modeled as well. Specifications of systems created using the framework can be simulated directly. The framework proposes rich means for analysis of simulation results, based on the technique described in [1]. An application of the framework in the context of a subway evacuation case is described in the paper. In particular, results are presented of the analysis how the penetration rate of ambient devices and the quality of knowledge of human and ambient agents influence the evacuation dynamics. The simulation model developed is based on a set of validated assumptions, provided by domain experts. This increases the confidence in obtained simulation results. The paper is organized as follows. The evacuation case is introduced briefly in Section 2. Then, the modeling approach is presented in Section 3 with a particular focus on decision making of agents. After that the analysis results for the case are described in Section 4. Section 5 concludes the paper.

2 Case Study

In this section a subway evacuation case is described investigated in the frames of the SOCIONICAL project:

An explosion occurred in a train situated in an underground tunnel between two stations. The explosion was followed by fire and smoke, which have been spreading throughout the tunnel. Because of the blast some doors are damaged, so people are able to escape outside of the train through the formed openings. In other carriages all doors are closed and could not be opened manually. People from the first and last carriages have a possibility to escape through the driver's and back doors. Also, people from other carriages can get out of the train through one of the available exits using buffers between carriages. Some people believe that tracks might still be alive and prefer to stay in the train. The train driver puts much effort to release and guide people to exits of the tunnels. As the driver acts confidently and in a knowledgeable way, many evacuees have a high trust towards him.

Two types of ambient devices are distinguished in the case: centralized devices installed in tunnels (e.g., LCD screens and loudspeakers) and personal ambient devices carried by human agents (e.g., PDA's, cell phones). It is assumed that any device can contact another device within its covering radius. Furthermore, a human agent possessing a personal device is able to exchange information with it. Also, centralized devices provide information to human agents in their neighborhood. Both types of devices provide two types of information used in decision making by agents: (1) information about currently possible options to move (i.e., paths to exits on the subway map) and (2) information that can be used by agents to evaluate their options: (a) the availability of locations along different paths, and (b) the degree of danger based on the fire and smoke concentration at a location.

The train driver has a personal device which is able to communicate with the Network Control Center and to gain the most recent information from there. Some percentage of the passengers in each carriage possess personal ambient devices. Every carriage is equipped with wireless communication points, through which devices of the driver and passengers can communicate with each other. Some communication points are broken.

Two particular research questions are addressed in this case study: (I) How does the penetration rate of ambient devices in the population of agents influence the evacuation speed? (II) How does the completeness and correctness of knowledge of agents influence the evacuation effectiveness? To answer these questions simulation has been performed, which will be considered in Section 4.

3 Modeling Approach

To specify dynamic properties of a system, the order-sorted predicate logic-based language called LEADSTO is used [2]. Dynamics in LEADSTO is represented as evolution of states over time. A state is characterized by a set of properties that do or do not hold at a certain point in time. To specify state properties for system components, ontologies are used which are defined by a number of sorts, sorted constants, variables, functions and predicates.

LEADSTO enables modeling of direct temporal dependencies between two state properties in successive states, also called dynamic properties. A specification of dynamic properties in LEADSTO is executable and can be depicted graphically. The format is defined as follows. Let α and β be state properties of the form conjunction of atoms or negations of atoms, and e, f, g, h non-negative real numbers. In the LEADSTO language the notation $\alpha \twoheadrightarrow_{e,f,g,h} \beta$ means: if state property α holds for a certain time interval with duration g, then after some delay (between e and f) state property β will hold for a certain time interval of length h. For example, the property 'if agent A observes fire during 1 time unit, then after that A will run away from the fire during 1 time unit' is formalized as:

$$observation\_result(fire) \twoheadrightarrow_{0,0,1,1} performed(runs\_away\_from\_fire)$$

From the external perspective the behavior of an agent is specified by dynamic relations between agent's input and output states, corresponding to interaction with other agents and with the environment in a multi-agent system. An agent perceives information by observation and communication and generates output in the form of communication or actions.

From the internal perspective the behavior of an agent is characterized by causal relations between internal states of the agent, based on which externally observable behavioral patterns are generated. The externally observable behavior of an agent is determined largely by the agent's decision making. The process of decision making in a highly dynamic environment (e.g., during an evacuation) is influenced greatly by information provided to the agent by diverse information sources. The higher the agent's trust in an information source the more the agent is apt to accept information from this source and to adapt its beliefs using this information. Due to the space limitations only an informal description of the cognitive model is provided. For a complete formal description we refer to [3].

Beliefs of an agent are time-labeled internal representations created based on communication and observation results received by the agent. Beliefs are specified using the function $belief(p : STATPROP, v : VALUE, t : TIME)$, here p is the content of the belief and v is the degree of confidence of the agent that the belief is true from the interval $[0,1]$ at time point t.

Trust is an attitude of an agent towards an information source (e.g., another human agent, an ambient device) that determines the extent to which information received by the agent from the source influences agent's belief(s). The trust to a source builds up over time based on the agent's experience with the source. In particular, when the agent has a positive (negative) experience with the source, the agent's trust to the source increases (decreases). For more details on modeling beliefs, trust and interaction between them we refer to [3].

It is often claimed that most beliefs of a human agent are emotionally loaded (see e.g., [6]); also trust of an agent to a source and decision making are influenced by emotions. In the proposed framework emotional influences on cognitive state (such as beliefs, trust and decision making states) are modeled by *body loops* (see [4]) over these cognitive states:

cognitive state → preparation for the induced bodily response → induced bodily response → sensing the bodily response → sensory representation of the bodily response → induced feeling

For more details on modeling of emotions we refer to [3].

Agents act according to the adopted plans. A *plan* is a (partially) ordered set of actions, required to be done to satisfy agent's goals. Plans are represented as beliefs and can be communicated between agents. For example, a part of plan $P1$ that specifies for an agent the sequence of locations l1, l2, l3 is formalized as $belief(follows\_after\_in(go\_to\_location\_from(l3, l2), go\_to\_location\_from(l2, l1), P1), 1)$. Plan descriptions vary in the degree of specificity. Abstract (or partial) plans contain high level actions that can be further refined into more specific actions. Furthermore, alternative ways of refinement of high level actions can be specified. In the evacuation scenario an agent may have several plans how to reach possible exits. These plans may be partial and contain movement actions between significant locations only (such as junction points, locations with elevators, stairs). There could be different paths between two significant positions.

An agent evaluates an option using the following criteria:

(1) *Desirability of an option*: to which extent the option brings to the satisfaction of the current desire(s). In the evacuation scenario the desirability of an option depends on the agent's beliefs about how close the agent's state will be to the state of being evacuated after the execution of the option.

(2) *Possibility to realize*: indicates the agent's confidence that the option can be realized. In the evacuation scenario it is calculated based on the confidence values of the agent's beliefs about the accessibility of locations in the plan and the degree of danger of each of these locations (e.g., amount of smoke and fire). These beliefs take into account the agent's representation of the environmental dynamics (e.g., about smoke spreading) and the agent's average speed.

(3) *The amount of effort*: how much effort needs to be invested to realize the option; depends on the skills, knowledge, physical and other characteristics available to the agent. In the emergency scenario the amount of effort is calculated based on the agent's beliefs about the distance to the exit.

To model an emotional influence on the process of evaluation of options, the value $eval_p$ for option p, calculated based on (1), (2) and (3) is provided as input for a body loop (for more details see [3]):

$preparation\_state(b, v1)$ & $body\_state\_for(b, option(p))$ & $belief(eval(p, v2), 1, t)$
& $feeling(b, v3))$ & $present\_time(t)$
$\twoheadrightarrow_{0,0,1,1}$ $preparation\_state(b, v1 + \gamma 1 \cdot (g(v2, v3, v4, \beta 1, \beta 8(t, t2), \beta 9, \beta 10) - v3))$,
where $g(v2, v3, \beta 1, \beta 2, \omega 1, \omega 2) = 1/(1 + e^{-\beta 1 \cdot (\omega 1 \cdot v2 + \omega 2 \cdot v3 - \beta 2)})$

After all available plan options have been evaluated, the agent chooses and commits to the one with the highest feeling value. After the agent has committed to a plan, it evaluates all possible refinements of the first action of the plan along the same evaluation criteria as for plans and chooses the one with the highest feeling. Note that the agent re-evaluates the available plan options continuously and can retract from its plan commitment at any time point.

4 Simulation and Analysis Results

To answer the research questions for the evacuation case posed in Section 2, simulation of a number of scenario variants described in the upper part of Table 1 was performed. In the simulation the following setup was used: The train comprises 4 coaches. An explosion occurred in the last coach, which is close to Exit A. Exit B is further away from the train, however the path to that exit is free of smoke at the beginning of each simulation. The train is populated in random manner with 20 heterogeneous human agents (5 per coach) of different types represented equally in the agent population: conservative extrovert/introvert, neutral, flexible extrovert/extrovert, neurotic flexible extrovert (for more details see [3]). In the scenario's with ambient devices a centralized ambient device is located in the tunnel on the way to Exit B. It is assumed that human agents do not know each other; initially the trust relations between human agents are neutral (0.5). An exception is the train driver agent, who is initially highly trusted (0.9) by other agents. Initially human agents have average trust in their personal ambient devices (0.6). The simulation time is 24 minutes. The obtained simulation traces describing the evacuation of agents over time are visualized in Figure 1.

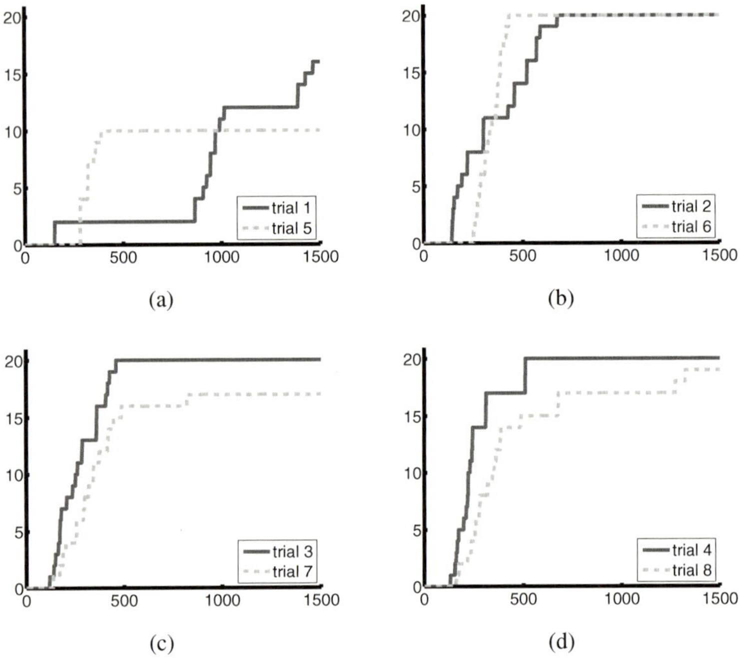

Fig. 1 The dynamics of evacuation in the simulation trials 1-8; the horizontal axis is time in seconds and the vertical axis is the number of evacuated human agents

Following [1], the dynamics of a simulation model can be studied by specifying dynamic statements in terms of temporal logical expressions, and automatically verifying these statements against simulation traces. For formalize such temporal statements the Temporal Trace Language (TTL) [1] is used, of which LEADSTO is a sublanguage.

To investigate how the penetration rate of ambient devices influences the evacuation speed two properties were verified on the simulation traces:

P6 - Evacuation Time t

There is a time point $t1$ in trace γ on which all agents are evacuated
and all other time points $t2$ on which agents are evacuated are later than $t1$
and the time that passed between the start of the simulation and $t1$ is t.

P7 - Quicker Evacuation Time
If in trace $\gamma 1$ there are $x1\%$ correct agents, $y1\%$ incorrect agents, and $z1\%$ AmI devices, and in trace $\gamma 2$ there are $x2\%$ correct agents, $y2\%$ incorrect agents, and $z2\%$ AmI devices, then the total evacuation time in $\gamma 1$ is quicker than in $\gamma 2$.

The verification showed that P6(trial 1, 1500), P6(trial 4, 1500), P6(trial 5, 1500), P6(trial 7, 1500), P6(trial 8, 1500) are not satisfied.

Based on the verification of property P7 on all simulation traces the following conclusions were made:

Conclusion 1: In all simulation scenario's adding ambient devices accelerates the evacuation process (in comparison with the scenario's without ambient devices).

Conclusion 2: In the scenario's, in which the percentages of agents with incorrect knowledge and with correct knowledge are both small ($\leq 10\%$) in comparison with the number of agents with no knowledge (simulation trials 1-4), adding a small number of ambient devices (10%) is sufficient to accelerate the evacuation process significantly.

Conclusion 3: For the scenario's, in which the numbers of agents with correct knowledge and with incorrect knowledge are both high (simulation trials 5-8), exists a penetration rate threshold; below this threshold the speed of the evacuation is improved significantly in comparison with the scenario's without ambient devices, and above the threshold the evacuation slows down.

By the following property we investigated the relation between the quality of knowledge of agents and the evacuation effectiveness:

P9 - Knowledge Distance for Agent a
If in trace γ at time t
agent a has belief i_1 with strength x_1 and ... and a has belief i_n with strength x_n
and i_1 in the world has strength y_1 and ... and i_n in the world has strength y_n
then the knowledge distance for agent a is $\sqrt{\sum_{i=1}^{n} (x_i - y_i)^2/n}$.

Table 1 An overview of simulation trials 1-10 and the quality of knowledge of the evacuated agents in these trials; three categories of quality are distinguished: High: $d \in [0, 0.3)$, Average: $d \in [0.3, 0.7]$, and Low: $d \in (0.7, 1]$

Simulation trial / Characteristic	1	2	3	4	5	6	7	8
Penetration rate	0	10	50	100	0	10	50	100
% agents with correct beliefs	10	10	10	10	50	50	50	50
% agents with incorrect beliefs	0	0	0	0	50	50	50	50
# agents with *High* knowledge	1	6	7	7	7	13	8	11
# agents with *Average* knowledge	13	9	10	8	3	3	5	4
# agents with *Low* knowledge	2	5	3	5	0	4	4	4

The results of verification of this property for the evacuated agents are provided in the lower part of Table 1. As one can see from the table, almost in every simulation trial agents with low quality of knowledge were evacuated. Partially this result can be attributed to the structure of the world used in the simulation. Partially this is because some agents found their way to an exit by following other agents.

Note that the results presented are based completely on the obtained simulation traces and still require validation.

5 Conclusions

In this paper a novel, agent-based framework is proposed for declarative modeling and analysis of ambient intelligence systems. The framework extends the LEADSTO [2] / TTL [1] analysis framework with explicit mechanisms to represent Ambient Agents. An important advantage of the framework over existing approaches (cf [8], [7], [9]) is that it allows modeling detailed cognitive processes from which intelligent behavior of agents emerges. The automated analysis means provided by the framework allows detailed qualitative and quantitative investigation of the system behavior in different scenarios and contexts. For example, the presented case study enabled the modeller to investigate which types of evacuation patterns would occur in given scenarios. In the future more sophisticated types of agents will be considered, which realize diverse Theory of Mind aspects.

Acknowledgements. This research has been conducted as part of the FP7 ICT Future Enabling Technologies program of the European Commission under grant agreement No 231288 (SOCIONICAL). See `http://www.socionical.eu/` for details.

References

1. Bosse, T., Jonker, C.M., van der Meij, L., Sharpanskykh, A., Treur, J.: Specification and verification of dynamics in agent models. Int. J. Cooperative Inf. Syst. 18(1), 167–193 (2009)
2. Bosse, T., Jonker, C.M., van der Meij, L., Treur, J.: A language and environment for analysis of dynamics by simulation. International Journal of Artificial Intelligence Tools 16, 435–464 (2007)
3. Bosse, T., Sharpanskykh, A., Treur, J.: A formal cognitive model of an ambient agent. Technical Report TR-02-2010, Vrije Universiteit Amsterdam (2010), `http://www.few.vu.nl/~sharp/tr022010.pdf`
4. Damasio, A.R.: The Feeling of What Happens: Body and Emotion in the Making of Consciousness. Harcourt Brace & Company, New York (1999)
5. Davidsson, P.: Multi agent based simulation: Beyond social simulation. In: Moss, S., Davidsson, P. (eds.) MABS 2000. LNCS (LNAI), vol. 1979, pp. 97–107. Springer, Heidelberg (2001)
6. Janis, I.L., Mann, L.: Decision Making: A Psychological Analysis of Conflict, Choice and Commitment. Free Press, New York (1977)

7. Shirehjini, A.A.N., Klar, F.: 3dsim: rapid prototyping ambient intelligence. In: Proceedings of the sOc-EUSAI joint conference, pp. 303–307. ACM, New York (2005)
8. Satoh, I.: Software agents for ambient intelligence. In: Proceedings of IEEE International Conference on Systems, Man and Cybernetics, SMC 2004, pp. 1147–1150. IEEE, Los Alamitos (2004)
9. Stathis, K., Toni, F.: Ambient intelligence using kgp agents. In: Markopoulos, P., Eggen, B., Aarts, E., Crowley, J.L. (eds.) EUSAI 2004. LNCS, vol. 3295, pp. 351–362. Springer, Heidelberg (2004)
10. Weber, W., Rabaey, J., Aarts, E. (eds.): Social, Economic, and Ethical Implications of Ambient Intelligence and Ubiquitous Computing. Springer, Heidelberg (2005)

Autonomous and Context-Aware Scheduling for Public Displays Using Place-Based Tag Clouds

Fernando Reinaldo Ribeiro and Rui José

Abstract. Public displays are an increasingly ubiquitous element of our socio-digital landscape with a strong potential to become key building blocks for AmI. For that purpose, they should have the ability to dynamically and autonomously select content from web sources according to the relevance of their content to the continuous flow of social settings around the display. In this paper, we explore the creation of a public display system that evaluates the relevance of content from web sources and selects the most relevant content according to a dynamic tag cloud that incorporates static place definitions, but is also sensitive to the people around the display. We have developed and evaluate a dynamic content selection system based on those principles. The results show that place visitors recognize the sensitivity of the system to their demands and that a place tag cloud can provide an important element for the interpretation of place and for the combination of the interests expressed by the place owner and the multiple place visitors.

1 Introduction

In Ambient Intelligence (AmI), smart interfaces can perceive and react to people, sense ongoing human activities and proactively respond to them. Public displays are an increasingly ubiquitous element of our socio-digital landscape with a strong potential to become important building blocks for AmI. However, most public displays today are mere distribution points for pre-defined and centrally created content that is not very valued by their potential users and is often perceived as too

Fernando Reinaldo Ribeiro
Informatics Department, Polytechnic Institute of Castelo Branco, Portugal
e-mail: fribeiro@ipcb.pt

Rui José
Information Systems Department, University of Minho, Portugal
e-mail: rui@dsi.uminho.pt

J.C. Augusto et al. (Eds.): ISAmI 2010, AISC 72, pp. 131–138.
springerlink.com

institutional or dull [1]. The key reasons for this limited success revolve around the limited content made available or their total unawareness regarding the Human activities taking place at that same location. Likewise, the solution may be found in the ability to dynamically integrate content from the web and select content sources according to their relevance to the social context around the display.

In this paper, we describe the creation and evaluation of a public display system that autonomously selects from web sources the content deemed more relevant according to a dynamic place model that is sensitive to the people around the display. People can express their content preferences by specifying tags in their Bluetooth device name, as described in [2]. These tags, together with seed tags defined by a place owner constitute a tag cloud that is continuously reflecting the social setting around the display, being sensitive to immediate indications of interest and providing a balanced combination between the content suggestions expressed by multiple place visitors and those expressed by the place owner. This tag cloud is then used as the basis for an adaptive scheduling algorithm that addresses the specific requirements raised by content recommendation in public displays. The results show that place visitors recognize the sensitivity of the system to their demands and that a place tag cloud can provide an important element for the interpretation of place and for combining the dynamic set of interests expressed by multiple people.

2 Related Work

Determining what to present next is a central feature for any public display system. In commercial systems, this is typically based on a fixed scheduling cycle of pre-defined content. Adaptive scheduling alternatives have been explored that introduce sensibility of the schedule to some type of external variable. Proactive displays [3] recognize specific presences and display information from associated profiles. Context-aware informative display [4] and context sensitive public display for adaptive multi-user information visualization [5] also aim to provide people with relevant content in an opportune and personalized way. Dynamo [6] or BlueBoard [7] are examples of systems that give users direct control of the display and thus content selection is directly handled by users rather than by the system. BlueScreeen [8] bases its scheduling decisions on history observation of exposed Bluetooth devices, which were collected during the advertising cycles. Muller [9] proposes a system that uses a naïve Bayes classifier and information about history, current time, location, advert keywords and feedback to determine the best advert to display. These systems support unassisted and adaptive scheduling, but they employ customized scheduling algorithms based on users' profiles or content selected from a predefined list.

In its essence, the problem addressed in this work shares many objectives with recommendation systems, which are usually classified into three main categories: content-based, collaborative, and hybrid recommendation approaches [10]. A recommendation system is focused on estimating ratings for items that have not been seen by a user. This estimation is usually based on the ratings given by this user to other items or based on ratings given by other users with similar interests.

However, the specific scenario of recommending content from web sources on a public display raises specific challenges that break some of the assumptions we often find in recommendation algorithms. The first challenge is that in a public display interaction is much more limited. In many cases, the intended use is simply to be seen, which does not generate any information at all about how they are used, much less about the users interests or preferences. Also, because there is normally no intermediate stage between content being suggested and being presented, there are considerably higher risks associated with inappropriate recommendations.

The second challenge is how to recommend content for a place instead of a person. Rather than having a user profile, a public display needs a place profile that combines the preferences of the person managing the display, who we call the place owner, with the preferences of the multiple users that may use the display.

Finally, and because we are considering dynamic content from web sources, we must address the fact that the relevance of the respective information is likely to face considerable oscillations, ranging between rich recently updated information and strongly deprecated or even non existing information. Together, these characteristics raise specific recommendation timings that are not common in most recommendation problems. The notion of most relevant content may change very quickly because of changes in the content itself, changes in competing content or changes on the set of people being sensed. As a consequence, the adaptation processes at the various selection levels must all be very dynamic and frequently re-evaluate their selection decisions.

3 A Place-Based Content Scheduler

A key part of this work was the specification and implementation of a content scheduler, represented in Fig.1, that is able to select the information feeds deemed more relevant for the current social setting.

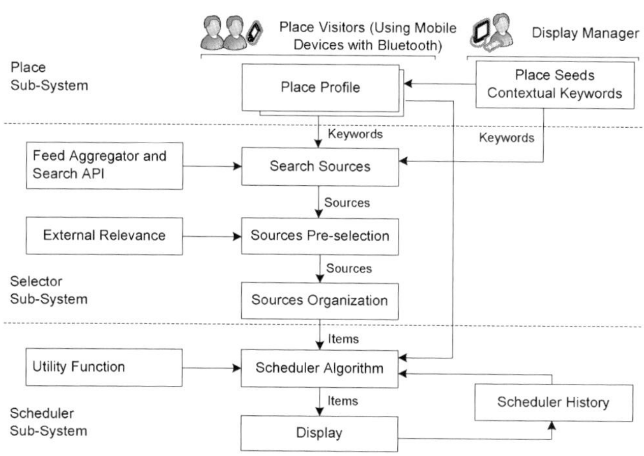

Fig. 1 System Architecture

There are three main components in this architecture: A place model that represents a dynamic and evolving view of the place in the form of a tag cloud, a selector that takes the keywords in the tag cloud and retrieves relevant content for each of those tags, and a context-aware scheduler that considers the weight and current presence of the represented tags to select which content is going to be shown next.

Place model. Place is a very broad concept, involving situated practices and a fluid interpretation of the surrounding social context. The model used in this system to represent place is based on a shared and historical view of the expressions of interest made by the people around the display. The model must also combine the expression of interests from a place owner, who is expected to exert stronger influence in the process of place making, and the multiple place visitors. Our approach to these issues is based on a place tag cloud. A tag cloud can combine multiple sources of content preferences, can easily be visualized to support interpretation and facilitates keyword-based content selection. The place tag cloud is first created by the place owner, who may define a set of seed words together with additional parameters that determine if those words always remain in the tag cloud or if they can be replaced by new words as they enter the tag cloud. A list of blocked keywords may also be specified as a simple filter for inappropriate content. Place visitors can also publish their own interests through *tag* commands in their Bluetooth device name [2]. The combination of multiple contributions from place visitors through time characterizes place as a seamless sequence of many social settings.

Selector. The selector sub-system is responsible for autonomously finding and selecting relevant content sources from the Internet. In this implementation, all the sources are dynamically selected and thus there are no predefined sources. The selection is based on the keywords from the place tag cloud and for each of the keywords the selector, selects the set of content considered more relevant. The selection within each keyword is based on source relevance. This is a non contextual measure and it is calculated as a function of two external indicators: the *rank* from *Alexa Traffic Rank* and the number of *subscribers* of each source. This gives an important marker that can help to improve the system utility, pre-selecting relevant sources and eliminating those sources that no one considers relevant.

Scheduler. The scheduler sub-system selects which of the tags in the place tag cloud will be used next for content presentation. For each tag there is information about the respective popularity and whether or not it is currently present, i.e. someone who is now around the display is announcing that tag. Additionally, there is also a waiting queue with the most recently selected tags. These tags are excluded and they are released again after a few selections. The algorithm selects the most popular tag among those who are present and not in the waiting queue, or if none is present, the most popular tag not in the waiting queue. When a tag is selected, the next step is to select which of the associated content items will be displayed next, using a multi-criteria utility function that considers multiple elements such as the length of the text, the content language, the number of links, and timeliness [11].

4 Evaluation and Discussion

The evaluation of this system was mostly based on user perception about the system behaviour rather than on its accuracy. It would always be subjective to set any comparison metrics for assessing the appropriateness of content to the social environment around the display. Therefore, we chose to focus on how users perceive the behaviour of the system within their own conceptual frame of the surrounding context. More specifically, our evaluation was focused on the following goals: the appropriateness of the place tag cloud as a shared and public model for the social setting around the display; the viability of tag indications as a simple interaction mechanism; the recognition by place visitors of the sensitivity of the system to their demands; and their overall perception that the content displayed is relevant for the place as expressed by the tag cloud.

4.1 Experimental work

To evaluate the recommender system we have conducted a 3 weeks experiment at the main entrance of the Information Systems Department at UMinho. This is mostly a transient place where students, teachers and staff walk by on their way to classes and labs, occasionally forming small groups. The prototype is composed by a large display that includes three panels (see figure 2).

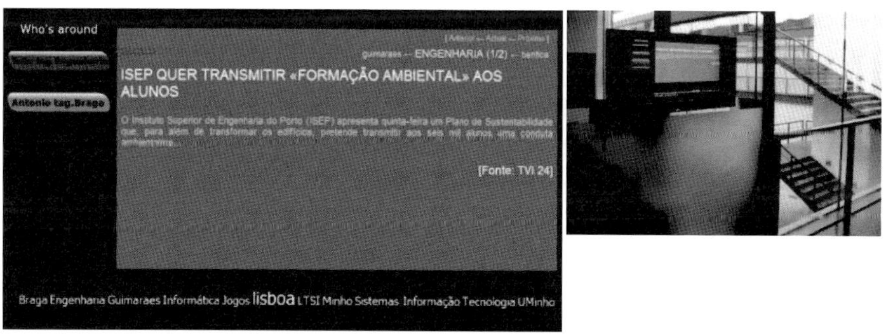

Fig. 2 a) Public display screenshot. b) Prototype image.

The left panel presents Bluetooth device names that are currently detected in the place. This is the *"who is around"* panel and essentially serves as a first level of presence recognition. People can add tags to their Bluetooth device name. The tag is added to the place tag cloud, if new, and marked as being present. Its popularity will steadily begin to increase while that presence remains. The place cloud is represented in the bottom panel. The size of the tags is proportional to their popularity in the place and keywords that are currently present are shown in yellow as an additional level of presence recognition and interaction feedback. The main panel presents content according to the current place profile. At its top, this panel displays the sequence of keywords being used for content selection, more specifically, the previous, the current and the next tags being used, thus facilitating the interpretation of the system behaviour by users.

For this experiment the display was configured to favour Portuguese content, to avoid presenting the same item every 10 presentations and to avoid present items from the same source every 5 items. The canonical view of the place was set by defining a number of seed tags for the place tag cloud associated with Informatics and Engineering topics as well as location related keywords associated with the town and region. During these 3 weeks we collected usage logs and conducted a total of 15 structured interviews with people at the place who had previously tried to use the system.

4.2 Results and Discussion

Table 1 summarises the usage statistics observed during the 3 weeks of the experiment between 9am and 8pm of the working days.

Table 1 Experiment statistical data

Distinct Bluetooth device names detected	349
Bluetooth devices detected	308
Total explicit interactions (tag commands across multiple sessions)	73
Distinct explicit interactions (unique tag commands)	31
Distinct Bluetooth devices with explicit interactions	23
Number of scheduled items	21616
Number of items selected from interactions originated tags	5567
Number of items selected from tags while those tags were present	360
Number of selected sources	128

The numbers show that there were 23 devices originating tags to the display. As a result 1,67% (360) of the scheduled items were presented as a direct reaction to tag publications while the respective user was present. This reflects the relatively limited number of people interacting with the system and above all the transient nature of the place, with most people just stopping for a very short interaction and then moving away. We can also observe that 25,7% of all schedulings were consequence of users interactions (with the remaining 74,3% being based on seed tags). This number is particularly interesting because it reflects the influence of the aggregate of interactions in the display behaviour even after the people making the interactions are gone. This number is strongly affected by the interaction patterns, but also by key behaviour parameters such as the decay level in the popularity of tags or the minimum popularity defined for seed tags. A quick decay will make the system more responsive to new arriving tags, but it will also fallback more quickly to the seed tags. A slow decay will tend to consolidate more commonly used tags, favouring an emerging view of place, but it will be much more difficult for new tags to reach the new level of popularity as those already in the tag cloud. This is therefore a key parameter in setting the responsiveness of the system and the balance between a pre-defined and a situated view of place.

To evaluate people's perception about the sensitivity of the system to their requests, uncovering opinions about the relevance of the content displayed, its timeliness and responsiveness to indications of interest, we also ask 15 place visitors who had used the systems to answers 6 groups of questions related with their experience. For each group there was also an open question where users are able to make their comments about the topic. Table 2 presents the results organized by groups of questions.

Table 2 Survey summary (group results)

Questions group	Good	Sufficient	Poor	Bad
Relevance of the content presented in the display	48,8%	41,9%	7,0%	2,3%
Influence of the interactions in the display behavior	30,2%	58,1%	9,3%	2,3%
Impact and acceptance	24,1%	58,6%	17,2%	0%
Handle with abusive behaviors	0%	83,3%	0%	17,2%
Interactions	32,5%	45,5%	15,6%	6,5%
Privacy	76,9%	0%	23,1%	0%

Overall, these results indicate that the perception of users in regard to the behaviour of the system was mainly positive. Some users, however, considered the system reaction to be slow as they seemed to expect an almost immediate change in content when defining new tags. While an immediate response was never our goal and even considering that setting the right expectations may be enough to change user perception regarding this issue, the relatively slow responsiveness may still represent a real limitation to the use of this approach in transient places.

5 Conclusions

In this work, we explored the usage of a dynamic scheduling model for content selection for situated public displays. This model is supported by a shared and public place profile in the form of a place tag cloud that combines the multiple interests of the place owner and place visitors. The positive results obtained during the evaluation suggest that this is a viable approach to the problem of selecting relevant content for a dynamic view of place. In particular, the visual nature of the tag cloud seems to facilitate the interpretation of the system behaviour in a way that influences positively the user perception, even when the selections are not perfect. Further research is needed to evaluate across multiple settings the ideal values for some of the system parameters. For example the decay of user-suggested tags affects responsiveness and also the balance between pre-defined and emerging notions of place, while the size of non-repetition queues affects the balance between content quality and diversity. Overall, these results suggest that this may be a valuable step towards the emergence of dynamic place profiles that match the social expectations and practices of their evolving social setting.

Acknowledgments. The first author was supported by a Portuguese Foundation for Science and Technology scholarship (SFRH/BD/31292/2006).

References

[1] Huang, E.M., Koster, A., Borchers, J.: Overcoming Assumptions and Uncovering Practices: When Does the Public Really Look at Public Displays? In: 6th International Conference on Pervasive Computing, Sydney, Australia (2009)

[2] José, R., Otero, N., Izadi, S., Harper, R.: Instant Places: Using Bluetooth for Situated Interaction in Public Displays. IEEE Pervasive Computing 7, 52–57 (2008)

[3] McDonald, D.W., McCarthy, J.F., Soroczak, S., Nguyen, D.H., Rashid, A.M.: Proactive Displays: Supporting Awareness in Fluid Social Environments. ACM Transactions on Computer-Human Interaction 14 (2008)

[4] Zhu, M., Zhang, D., Zhang, J., Lim, B.Y.: Context-Aware Informative Display. In: International Conference on Multimedia and Expo., Beijing (2007)

[5] Morales-Aranda, A.H., Mayora-Ibarra, O.: A Context Sensitive Public Display for Adaptive Multi-User Information Visualization. In: Third International Conference on Autonomic and Autonomous Systems, Athens, Greece (2007)

[6] Izadi, S., Brignull, H., Rodden, T., Rogers, Y., Underwood, M.: Dynamo: A public interactive surface supporting the cooperative sharing and exchange of media. In: Symposium on User Interface Software and Technology, Vancouver, Canada (2003)

[7] Russell, D.M.: Large interactive public displays: Use patterns, support patterns, community patterns. In: Public and Situated Displays: Social and Interactional Aspects of Shared Display Technologies, Kluwer Academic Publishers, Dordrecht (2003)

[8] Payne, T., David, E., Jennings, N.R., Sharifi, M.: Auction Mechanisms for Efficient Advertisement Selection on Public Displays. In: European Conference on Artificial Intelligence, Riva del Garda, Italy (2006)

[9] Müller, J., Schlottmann, A., Krüger, A.: Self-optimizing Digital Signage Advertising. In: Adjunct Proceedings of Ubicomp., Innsbruck (2007)

[10] Adomavicius, G., Tuzhilin, A.: Toward the Next Generation of Recommender Systems: A Survey of the State-of-the-Art and Possible Extensions. IEEE Transactions on Knowledge and Data Engineering 17 (2005)

[11] Ribeiro, F.R., José, R.: Timeliness for dynamic source selection in situated public displays. In: 5th Int. Conference on Web Information Systems and Technologies, Lisbon (2009)

Semantic and Implicit HCI: A Step Forward AmI Delivery at Home for Patient Experience

Ana Isabel Calvo-Alcalde, Juan José Andrés-Gutiérrez, Jesús Vegas, Rafael Cano-Parra, and Esteban Pérez-Castrejón

Abstract. Spanish population over 65 years old will soon be an important fraction. These data show the need for some helping technologies that make possible to deal with this scenario. This research aims to explore Ambient Intelligence and Semantic Web technologies to allow personal health monitoring and user interactions at home as transparent as possible. To achieve this goal a Multi-Agent System is proposed, focusing on knowledge representation. The ontology has been defined according to current standards in order to improve the interoperability of the whole system, and its possible communications with other external systems. Furthermore, the defined language supports the semantic reasoning process for agent's real time decision making according to changing circumstances. This proposal can be understood as a step forward patient experience, in the sense of how environment information and user`s behaviors are taken into account to provide and manage suitable services.

Keywords: Ambient Intelligence, Multi-Agent Systems, e-Health, ontology, Ambient Assisted Living, semantic reasoning, implicit Human-Computer Interactions, patient experience.

Ana Isabel Calvo-Alcalde · Jesús Vegas
Computer Science Department
University of Valladolid, Escuela Técnica Superior de Ingeniería Informática (E.T.S.I.I.)
Campus Miguel Delibes s/n, 47001, Valladolid, Spain
e-mail: anaisabel.calvo@alumnos.uva.es, jvegas@infor.uva.es

Juan José Andrés-Gutiérrez · Rafael Cano-Parra · Esteban Pérez-Castrejón
Telefónica R&D
Parque tecnológico de Boecillo, Boecillo, Valladolid, Spain
e-mail: {jjangu,rafacp,esteban}@tid.es

J.C. Augusto et al. (Eds.): ISAmI 2010, AISC 72, pp. 139–147.
springerlink.com

1 Introduction

Nowadays society is involved in a new design trend aimed to increase user experience and services or products usability. For many years the best service organisations in the private sector have been aware of the need to manage the "customer service experience" with the same rigour as they bring to their functional and operational components. This new movement designates *experience* as 'how well people understand it, how well feel about it while there are using it, how well it serves its purpose, and how well it fits into the context in which they are using it'. Related to this, a new specific concept comes up in the medical environment. It refers to patient satisfaction due to the knowledge of these user-centred principles for a better service and a better *patient experience* [1].

Although the older population has a large buying power and ageing is becoming a global phenomenon, the *market of ICT, Information and Communication Technologies, for ageing* well in the information society is still on its nascent phase [2]. The social challenge is to keep the costs of healthcare systems under control while, at the same time, maintaining a high quality service. eHealth applications, based on ICT products and services, are expected to contribute to addressing this challenge minimizing the technological barriers related with older population. The technological standards for the Digital Home play a key role and can contributes providing not only entertainment services but also Ambient Assisted Living (AAL) services for the elderly at home [3].

Ambient Intelligence [4] is a forward-looking approach, in a near temporal horizon, in which people will live surrounded by transparent technology embedded in their environment, accessible whenever users need and through simple and natural interactions (addressed to every sensory senses) adapted to their preferences and context sensitive. Furthermore, Ambient Intelligence in the context of eHealth could be more than maintaining and improving existing health care systems. It could support a paradigm shift in health care delivery by focusing on the autonomous citizen and independent living [5]. In this sense it is worth to point out how these applications can also take advantage of implicit Human-Computer Interactions, HCI, in order to minimize the technological barriers related with older population. The system manages a context image, made of data from the hybrid sensor network, without user's awareness as possible.

This research aims to present a system focusing on AmI, implicit HCI and the Semantic capabilities supported by the intelligence of the architecture and, specifically, by the defined *Multi-Agent System MAS* [6]. The second section presents a general overview of the state of the art. Then, third section presents an analysis of the proposed MAS and the ontology defined as a common language. The system technological architecture is shown in section four, and finally, sixth section reports some conclusions and future works.

2 State-of-the-Art

Nowadays there is a huge amount of technological devices available on the market and people are used to interact with them to carry out daily task. In the scope of Digital Home we can conclude that many of them share several features. For example, many of them are designed to be more and more portable and trying to allow user interaction through touch screens. This market evolution creates new opportunities to offer value added services focusing on final users and enabling AmI environments. Devices that have perceptional capabilities will start the shift from explicit HCI towards a more implicit interaction with machines [7].

O'Hare previous researches [8] [9], advocate the use of software agents as a key enabler in the delivery of Ambient Intelligence. This technology is specifically useful for technology integration into everyday tasks, allowing users' interaction with their surrounding environment simpler and more intuitive. Also, MAS have been chosen as a common infrastructure adopted by multimodal HCI systems, pointing out that Multi-agent architectures provide essential infrastructure for coordinating the many complex modules needed to implement multimodal system processing and permit this to be done in a distributed manner [10].

On the other hand, the advantages of using Semantic Web languages for context information modelling have been presented in several studies [11] [12]. Some interesting reasons are that ontologies specify a common language for all the participants involved and they can support reasoning processes. The first work in which Web Ontology Language OWL [13] was used for element description in a context-aware system was CoBrA [14]. It is an agent based architecture which uses ontologies to model data and uses a rule based inference engine. COBRA, the first standard ontology for context, was presented. It reuses the vocabulary defined in Standard Ontology for Ubiquitous and Pervasive Applications SOUPA [15].

Advanced healthcare applications have been under prototyping some years by now; their usually imply the target user to wear sensors, and their main objective is to anticipate or detect health risks. For example, Kang et al. [16] propose a wearable sensor system which measures the biological functions of a person to provide remote health monitoring and self health check. Korel and Kao [17] also monitor the vital signs and combine them with other context info such as person's condition, in order to detect alarming physical states.

Moreover, several prototypes encompass the functionalities mentioned above: Rentto et al. [18], have developed a prototype of a smart home that integrates the context information from health monitoring devices and the information from the home appliances. Becker et al. [19] describe the amiCa project which supports monitoring of daily liquid and food intakes, location tracking and fall detection. Haigh et al. [20] developed an agent-based monitoring and support system that issues reminders, alerts and notifications, generates summary reports of patient's behaviour and provides an overview of person's state and medication compliance.

Although some of these previous systems have been evaluated for a reduced number of users, the review of them for home care domain indicates that this application area is still in its early stage. The proposed system is based, since its early state, in several technologies with the purpose of minimizing the technologicalbarriers usually present in this kind of systems. Moreover, the definition of an ontology based on standards increases the whole platform integration capability with other external systems, simplifying the frequent information exchange processes when patients are monitored at home. Finally, it is worth to point out how the platform takes advantage of semantic techniques for knowledge inference using the implicit context information from user's daily activities at home.

3 Framework AmI Based on Semantic Web Technologies

3.1 Multi-Agent System Integration in Digital Home Environment

A Home consists of heterogeneous devices with different communications capabilities (WiFi, Bluetooth, etc.), that is to say, a wide ecosystem of devices with sensors that enable them to identify and locate users (RfID, Radio Frequency Identification, readers), to measure vital constants (pulsioxymeter, scales, etc.) and to obtain environmental data (meteorological station). In order to incorporate these devices and, of course, to put their functionalities at the disposal of a Home, a new concept must be introduced: that of *service discovery*. This is possible using *Universal Plug and Play UPnP* [21] which, (1) provides automatic discovery of any devices connected to the Home network, (2) allows zero-configuration networking, (3) is independent of the kind of devices and networks, (4) uses standard protocols and (5) offers easy system extensibility. Once devices are able to put their services at the disposal of the Home, it is needed another component to manage and organize the intelligence provided: the global platform, Figure 1.

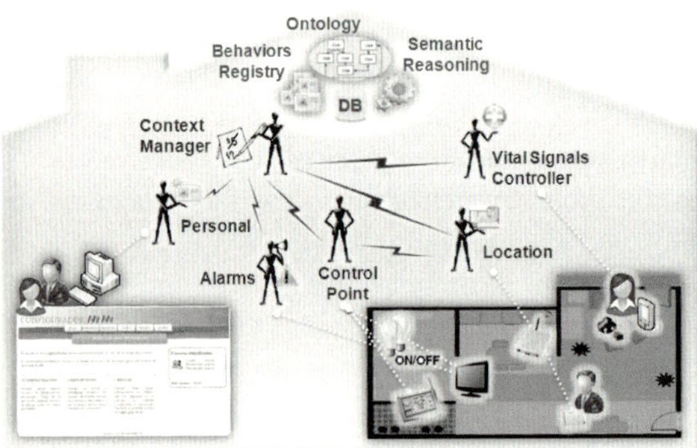

Fig. 1 Multi-Agent System general overview

A MAS has been defined which consists of autonomous intelligent agents which are able to communicate, can perceive the environment around them and coordinate their activities in order to achieve the goals they were designed to. This system allows users explicit interactions to create personal rules in order to define the system behaviour, that is, to associate the various events with the actions that should be carried out. On the other hand, the system is in charge of scanning the home network, integrating the found devices in home network. Using the environment's information and any personal rules defined for the user, the action that should be performed can be dynamically decided.

3.2 Ontology Based on Standards

The whole system capability to reason and its ability to be satisfactory adapted to environment changing circumstances depends on the *information model and its suitable management.* Therefore, data representation must be defined according to a model that represents them, as far as possible. Moreover, if it is based on standards, the platform will be able to exchange information with other external systems using the same language, making easier the integration processes. The data modelling proposal has been carried out applying the 'reutilization rule' and the following ontologies have been used during the definition of the current frame ontology, in order to allow its future improvement and extension.

Table 1 Related standard OWL ontologies for system interoperability

Ontology	Description
SOUPA	*Standard Ontology for Ubiquitous and Pervasive Applications*
(Core and Extensions)	Consists of nine ontology documents which define vocabularies for describing: person contact information, beliefs, desires, and intentions of an agent, actions, policies, time, space, and events.
MoGATU BDI	Describes an abstract model for representing human users and software intelligent agents: actions, believes, desires, intentions, and goals.
FOAF	*Friend Of A Friend*
	Personal information and relationships: persons, agents, projects, groups, documents, images, online accounts…
DAML-Time	Temporal concepts and properties to any formalization of time.
OpenCyc - OpenGIS	Vocabularies for expressing spatial relations (location).
Rei Policy	Deontic concepts (i.e., rights, prohibitions, obligations, and dispensations) for specifying and reasoning about security access control rules.
Basic Geo	Namespace for representing lat(itude), long(itude) and other information about spatially-located things: spacialThing and point.
CODONT	*Context-Driven Adaptation of Mobile Services*
(CoDAMos)	The system enables any service to detect changes in the user context and to dynamically adapt the services to this new context: users, services, platforms and environmental conditions.

The following elements support the semantic reasoning process:

- *The defined OWL ontology:* Represents the context and allows the reutilization of related standard ontologies, like the ones listed in Table 1. In order to provide eHealth services, the core data model is designed to cover some basic entities, such as patients, paramedics, doctors and medical personnel, and a description of the settings where services are provided in terms of the available devices (sensors and actuators).
- *Semantic Web Rule Language, SWRL [22], rules base:* It is a W3C recommendation with a higher expressive capability than OWL useful to represent context complex situations. It has been chosen to specify the dependencies between the relationships of the OWL individuals, and to reason and infer new knowledge about them. Moreover, the behaviors rules have been translated to this language to allow semantic reasoning for the inference process.
- *The Multi-Agent System:* It will be in charge of the ontology management, and will decide dynamically the suitable actions that must be carried out, taking into account the new inferred knowledge.

4 Technological Design to Support AAL Services at Home

Several enabling technologies have been presented with the purpose of designing a system to support users to carry out their everyday life activities in a natural way using the environment information. The proposed technological design allows the development of an *open and scalable system*, in which new agents and devices can be integrated easily. This quality is of special interest in digital home environment; due to the high number of heterogeneous devices that frequently came onto the market. Moreover, third-party components usually distributed as OSGi, Open Services Gateway Initiatives, [23] bundles can be also integrated *increasing the overall supported functionality.*

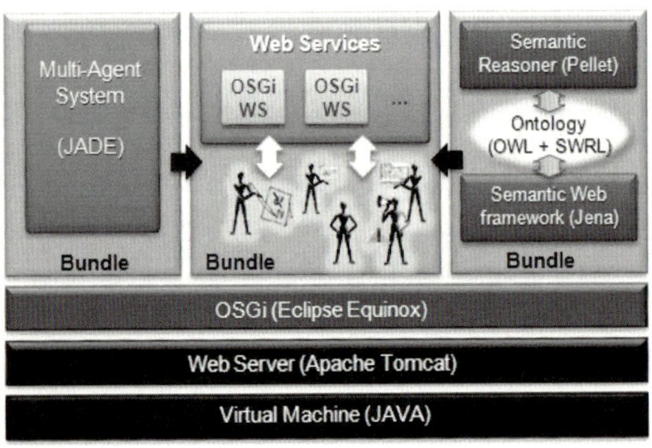

Fig. 2 Technological architecture

The general philosophy that promotes the practical implementation of the AmI framework has been SOA, Services Oriented Architecture, in particular, OSGi platform has been chosen. The MAS has been developed in the highest layer of OSGi, supporting the provided OSGi web services. Figure 2 shows the technological architecture and all the elements involved in the solution.

Therefore the MAS has been implemented in Java using the JADE agent platform [24], base on FIPA standard. On the other hand the OWL ontology has been designed and developed with Protégé-3.3.1, specifically, with its extension Protégé-OWL editor. Then, BeanGenerator plug-in has been used as the integration gateway to generates Java files, the ones which later have been imported in JADE. Finally, the declarative rules have been represented in SWRL and Pellet [25], an Open Source OWL reasoner, has been used to support the semantic reasoning process.

5 Conclusions and Future Work

The presented initiative focuses on the extension of the time people can live in their preferred environment by increasing their autonomy and self-confidence. All the technologies have been chosen in order to decrease the technological barriers usually related with elderly individuals.

With this purpose the system take advantage of the implicit user´s interactions and tries to reduce to a minimum the amount of intervention on the patient's part; in fact the patient must just put the sensors in place, and the system will carry out the transmission of data to the home environment, and to the hospital server. The project is specially intended for people who need constant monitoring, since it allows sanitary personnel to be informed when any anomalous situation arise.

It has been proposed a multi-agent system which takes advantage of current Semantic Web techniques to manage the context, pointed out as the main element during the reasoning processes of the AmI framework. Moreover, it has been presented how these capabilities can be useful in order to improve the interoperability of the whole platform with other external systems, and finally, how users will experience a better e-health services provision at their homes.

This research opens up several future lines: (1) The incorporation of data mining techniques for inference process improvement, (2) Incorporation of learning capability from past situations, (3) The integration of DLNA devices, (4) Development of an adaptive remote control for different devices and (5) To carry out a usability study of the user interface to improve the explicit HCI.

References

1. Bate, P., Robert, G., Maher, L.: Bringing user experience to healthcare improvement: The Concepts, Methods and Practices of Experience-based Design. Radcliffe Publishing (2007)
2. European Commission, Ageing well in the Information Society. An i2010 Initiative, Action Plan on Information and Communication Technologies and Ageing (2007)

3. Pérez-Castrejón, E., Andrés-Gutiérrez, J.J.: AAL and the Mainstream of Digital Home. In: Proceedings of the 10th International Work-Conference on Artificial Neural Networks: Part I (2009)
4. Weber, W., Rabaey, J.M., Aarts, E.: Ambient Intelligence (2005)
5. Aarts, E.H.L., Encarnação, J.L.: True Visions: The Emergence of AmI. Springer, Heidelberg (2009)
6. Stiglic, G., Verlic, M., Kokol, P.: Software Agents. Wiley encyclopedia of biomedical engineering. John Wiley & Sons, Chichester (2006)
7. Schmidt, A.: Implicit Human Computer Interaction Through Context. Personal Technologies 4 (2000)
8. O´Hare, G.M., O´Grady, M.J., Keegan, S., O´Kane, D., Tynan, R., Marsh, D.: Intelligent Agile Agents: Active Enablers for Ambient Intelligence. In: AISD, Vienna (2004)
9. Vall´ee, M., Ramparany, F., Vercouter, L.: A multi-agent system for dynamic service composition in ambient intelligence environments. In: Proceedings of the 3th International Conference on Pervasive Computing, Munich, Germany (2005)
10. Jaimes, A., Sebe, N.: Multimodal Human Computer Interaction: A Survey. In: IEEE International Workshop on Human Computer Interaction (2005)
11. Gu, T., Wang, X.H., Pung, H.K., Zhang, D.Q.: An Ontology based Context Model in Intelligent Environments. In: Proceedings of Communication Networks and Distributed Systems Modeling and Simulation Conference, San Diego, California, USA (2004)
12. Chen, H., Finin, T., Joshi, A.: Semantic Web in the Context Broker Architecture. In: Proceedings of 2nd IEEE International Conference on Pervasive Computing and Communications (2004)
13. OWL, `http://www.w3.org/TR/owl-features/` (accessed January 12, 2010)
14. Chen, H.: An Intelligent Broker Architecture for Pervasive Context-Aware Systems, PhdThesis, University of Maryland, Baltimore County (2004)
15. Chen, H., Finin, T., Joshi, A.: The SOUPA ontology for pervasive computing. In: Ontologies for Agents: Theory and Experiences, Springer, Heidelberg (2005)
16. Kang, D.O., Lee, H.J., Ko, E.J., Kang, K., Lee, J.: A Wearable Context Aware System for Ubiquitous Healthcare. In: Proceedings of the 28th IEEE. EMBS Annual I. C., USA (2008)
17. Korel, B.T., Koo, S.G.M.: Addressing Context Awareness Techniques in Body Sensor Networks. In: Proceedings of 21st International Conference on Advanced Information Networking and Applications Workshops, vol. 2, Niagara Falls, Canada (2007)
18. Rentto, K., Korhonen, I., Vaatanen, A., Pekkarinen, L., Tuomisto, T., Cluitmans, L., Lappalainen, R.: Users' Preferences for Ubiquitous Computing Applications at Home. In: First European Symposium on Ambient Intelligence, Veldhoven, The Netherlands (2003)
19. Becker, M., Werkman, E., Anastasopoulos, M., Kleinberger, T.: Approaching Ambient Intelligent Home Care System. In: Pervasive Health Conference and Workshops, Innsbruck (2006)
20. Haigh, K.Z., Kiff, L.M., Myers, J., Guralnik, V., Geib, C.W., Phelps, J., Wagner, T.: The Independent LifeStyle AssistantTM. In: 16th Innovative Apps of Artificial Intelligence C (2004)
21. Kim, D., Lee, K., Song, H., Sung, J.: UPnP-Based Sensor Network Management Architecture. In: Second International Conference on Mobile Computing and Ubiquitous Networking (2005)

22. SWRL, http://www.w3.org/Submission/SWRL/ (accessed January 10, 2010)
23. OSGi, Open Service Gateway Initiative, http://www.osgi.org (accessed February 1, 2010)
24. JADE, Java Agent DEvelopment Framework, http://jade.tilab.com (accessed January 1, 2010)
25. Pellet, http://clarkparsia.com/pellet (accessed January 10, 2010)

Domotic Hardware Infrastructure in PERSONA Project

Vicenç Soler, Andrés Peñalver, Simone Zuffanelli, Jordi Roig, and Jordi Aguiló

The domotic hardware infrastructure of PERSONA project is described. PERSONA aims to help elderly people at home, and this paper shows the hardware selection process and describes the specific hardware development carried out to minimize both the final price of the product and the visual impact in a real home. To do it, three generic types of PCB, that interfaces all the sensors and actuators with the communication transceiver, have been designed and implemented in order to minimize production costs. Also other specific devices have been carried out to cover the specifications of functionality of the project.

1 Introduction

PERSONA EU Project (Perceptive Spaces prOmoting iNdependent Aging)[1] is an EU-funded research project (FP6) that started in 2007. It belongs to AAL (Ambient Assisted Living) and aims to help elderly people at home in the loose of skills due to the normal aging process. This help is carried out by providing a set of services that includes the development of technology (hardware and software). The main technology involved in this IP EU project involves the development of a specific Middleware, a specific Communications System (also inside and outside home) and a Hardware Infrastructure [1].

PERSONA EU project is divided in 4 main areas: business (tries to find a place in market for the outcome of the project), social (specifies the needs of the elderly people), technical (implements the solution) and ethical.

Vicenç Soler · Andrés Peñalver · Simone Zuffanelli · Jordi Roig · Jordi Aguiló
Dept. Microelectrònica i Sistemes Electrònics, Universitat Autònoma de Barcelona, 08193 Bellaterra, Spain

Vicenç Soler · Jordi Aguiló
Centro CIBER-BBN (Bioengineering, Biomaterials and Nanomedicine), Campus UAB, 08193 Bellaterra, Spain

[1] Work funded by the EU FP6 projects PERSONA (contract N. IST-045459).

J.C. Augusto et al. (Eds.): ISAmI 2010, AISC 72, pp. 149–155.
springerlink.com

This document just faces the work developed in the areas of Sensors and Actuators, that is, the domotic hardware infrastructure of PERSONA EU Project.

Thus, the main work issued in this part of the project consists in finding and developing appropriate hardware (find devices, develop boards) in order to fit the specifications proposed in the Social are of this project, taking into account that the project should be installed in real homes.

The hardware involved in this project belongs to different classes, and mainly, we have identified 5 types of devices:

- Sensors: security and environmental sensors.
- Actuators: security and environmental actuators.
- User Devices: multimedia devices information (video, audio, graphics, etc.) as TV, screens, speakers, etc.; and Data devices: the use of more information than multimedia (Personal Mobile Devices, computers, IP cameras, servers, etc.).
- Specific Devices (Health care devices and other specific devices).
- Smart Textiles.

This document reflects the necessity of installing a set of devices in an existing and usually old flat. So, there are some physical aspects to be taken into consideration, as there live elderly people (usually just one), where the environment should not be changed radically. Thus, PERSONA should avoid installing wires and placing wires or other objects on the ground and on the walls for both not increasing the probability of falling down and not changing the environment where the elderly lives, when possible.

Following this idea, the project should focus its efforts in placing wireless devices. Wireless devices not only from the point of view of their communications, but also from the point of view of their power supply. Thus, they should have a long life battery included in them.

Another key aspect to be taken into consideration is the final cost of the project, from the point of view of the user, since it is expected that PERSONA will be a final product in the market. To do it, we decided to create a set of PCB to connect the sensor/actuator and a transceiver (together with a microcontroller) that communicates the data to the PERSONA Central System. As homes have different sizes, walls, etc. Zigbee was chosen as the communication system, since it allows multihop communications by creating a mesh network, and it has low power consumption (from now on, the set sensor/actuator + PCB + microcontroller + transceiver is called Device). This solution had very good results respecting the final price since there exist a wide quantity of devices with no communication issues, cheaper and allowing that the specific communication protocol of the project could be programmed without being constrained by any device provider. Therefore, we chose (when possible) totally wireless devices (also linked with the previous aspect) to decrease the price of an installation, and trying to choose always cheaper devices in market.

That was the most important issue concerning hardware respecting other similar EU projects as, for instance, AMIGO[2], ASK-IT[3] and SOPRANO[4]: it was needed to develop new hardware that allows a low-cost integration of generic devices (usually sensors/actuators) found in market in the same communication network.

The overall system has not still installed in a real home, and for that reason we cannot report feedback from real users, but it has been tested in a Living Lab with successful results from users using it during several hours. We also expect good results in real homes. The chosen cities are San Vicente de Raspeig (Alicante, Spain) and Odense (Denmark).

This document is structured in three chapters: an introduction, the list of devices used and the PCB designed.

2 List of Devices

A PERSONA Device is composed by a PERSONA ZigBee communication system (transceiver + microcontroller) and a sensor/actuator. The sensor/actuator can be a commercial one or it has been created specially for this project, but the most majority are commercial, since its final price is lower.

The final list of sensors, actuators and specific designs is exposed in the following subsections.

2.1 Security Sensors

- **Smoke Sensor:** optical on/off sensor that detect if there is smoke.
- **Carbon Monoxide Sensor (CO):** on/off sensor that detects CO.
- **Gas Sensor:** on/off sensor that detects gas.

2.2 Environmental Sensors

- **Motion Detector:** on/off PIR sensor that detects motion.
- **Magnetic Contact Transmitter (Door, Window & Fridge):** Passive sensor that detects if the entry home door or a window or the fridge is open/close.
- **Faucet State Sensor.** On/off magnetic flow switch.
- **Home Appliances State Sensor:** It detects if some device is switched on. It is a plug interface of the device to the power grid. For instance, it is used to not forget any device switched on when sleeping.
- **Temperature/Humidity Sensor:** Digital sensor that provides the temperature and humidity inside home.
- **Water Detector (floor):** Infrared sensor that detects flooding.
- **Glass Break Detector:** Vibration sensor that detects if a window is broken.

2.3 Actuators

- **Switch on/off any device:** Interface of a It switches on/off a device. It is a plug interface of the device to the power grid.

- **Controllable Heater System:** Controls electronically the heater system of a home by applying voltage to it.
- **Blinds Actuator:** Motor that open/close blinds.
- **Curtain Actuator:** Motor that open/close curtains.
- **Water Actuator:** Motor that opens/closes the general faucet of the home.
- **Siren:** It activates a sound alarm.
- **Light Actuator & Detector:** It is a dimmer that controls any light at home. It can switch on/off any alight and provide its intensity.

2.4 Specific Designs

- **Electronic Weight:** electronic weight adaptable to wireless communications.
- **Bracelet:** it detects when the elderly falls down and the activity done with one hand (for controlling mobility exercises). It also will have included a tracking system to localize the elderly at home (this option can be used to control abnormal behaviour), a vibrator to inform about events and a panic button.
- **Beltclip:** it can measure the movement activity of the elderly and can control when the elderly falls down. Both the bracelet and the belt will complement each other the information regarding the falling down problem.

3 Designed PCB

To allow a low cost project, it was necessary to create a common platform to integrate all of them. To allow it, there were designed three kind of boards (PCB) that complied the general specifications of the majority of the sensors and actuators of a domotic project. That is, on/off sensors and actuators that may need a 230V power supply or battery supply.

This section is divided into two parts: the first one refers to generic PCB, valid for every domotic project, and the second one are specific PCB designs needed to carry out specific devices of this project.

3.1 Generic PCB

Mainly, the sensors and actuators, depending on their power consumption, can be categorized in two classes: low power consumption (batteries are enough) and 230V power consumption (typically motors, actuators and specific sensors). It is mandatory that low power consumption devices need to maintain the same batteries, at least, during one year.

There are three standard PERSONA ZigBee Boards: 230V Sensors for sensors that need wired supply, Portable sensors for sensor that has a low power consumption and Pure Relay Actuator for actuators. As actuators usually need much more power consumption than sensors, their generic PCB should not include the possibility of connecting a battery (just 230V).

3.1.1 230V Sensors Board

This group of devices is composed of sensors that, due to its power consumption and/or voltage supply type and value, can not work properly with batteries. The sensors can be active (needs of power supply) or passive, and most of them have relay outputs.

All the active sensors connected to this board operate at 12V DC.

The Microcontroller reads the information from the sensors, through Pull-up resistors connected to one of its input ports, and sends the collected data by means of its ZigBee Transceiver. In addition the board implements 2 analogic inputs at port 0.

Smoke Sensor , carbon Monoxide Sensor (CO), Gas Sensor, Motion Detector, Faucet State Sensor, Home Appliances State Sensor and Glass Break Detector are the sensors that use this PCB.

Schematic Design

The 230V Sensors Board has three parts: the 12V DC power supply (the usual sensors working voltage), the 3.3V DC power supply and the Microcontroller.

This board is directly connected to the 230V AC electric grid. The Power Supply unit is based on a 12V Transformer and supplies two power lines: a 3.3V (100mA) line to supply the microcontroller and a 12V (300mA) line to supply the sensors.

The 12V DC power supply is composed by a 12V transformer to reduce the voltage and to isolate the electric grid from the rest of the circuit, a rectifier bridge to transform the alternating current into a continuous one, a filter stage to eliminate all the current alternating components, and finally a regulation stage to improve the stability and output impedance of the power supply output signal. This signal is routed to an 8 pin clamp connector where the sensors are connected to obtain the supply they need. A fuse protects the board against peaks of current.

The 3.3V DC power supply is done by a regulator that, from 12V DC, provides a stable 3.3V DC signal to supply the microcontroller.

Finally, a 40-pin socket is included to connect the ZigBee transceiver. This socket connects it to the external sensors connector composed by 5 digital inputs and 2 analog inputs.

In addition to the above parts, the board includes an ON/OFF switch, a RESET Button connected to the microcontroller reset pin and three LEDs to show the state of the board: On/Off, Network status and Transmitting or Receiving.

3.1.2 Portable Sensors

This board is designed for the sensors that work properly with batteries and do not need to be connected permanently to the electric grid. These sensors are those that are supplied at 3.3V or do not need power supply (passive). All of them have relay outputs except the Temperature and Humidity sensor that communicates with the microcontroller using a digital protocol similar to the I2C.

Magnetic Contact Transmitter, Temperature/Humidity Sensor, Water Detector (floor) and the Electronic Weight are those that use this PCB.

However, in order to be prepared for new 'portable' sensors and/or actuators in the future, the board is designed to manage up to three 'portable' sensors, one 'portable' analog sensor and two 'portable' actuators. In this way, this board can be used as a universal interface to connect devices to the PERSONA System, with a 150mA maximum total current consumption.

The Microcontroller reads the information from the sensors, through Pull-up resistors connected to its Port 1 inputs, and sends the data by means of its ZigBee Transceiver.

Schematic Design

The Portable Sensors Board is divided into three parts, the 3.3V DC Batteries power supply, the sensors connection and the Microcontroller.

The board is power supply with two AAA 1.5V Batteries that provide 3V DC. The 3.3V DC signal is obtained by a charge pump voltage regulator that, from a minimum of 1.8V DC, provides a stable 3.3V DC signal to supply the microcontroller.

In addition to the above parts, the board includes an ON/OFF switch, a RESET Button connected to the microcontroller reset pin and three LEDS to show the state of the board.

3.1.3 Pure Relay Actuators

This group of devices is composed of actuators that work as a switch, that is, are managed through a relay. In general, these actuators are motors that manage blinds, doors, curtains, etc. and relays that switch on/of the home appliances.

The Microcontroller is the responsible for receiving the orders from the central system and to activate the actuators through the relays integrated in the PCBs.

All the actuators use this PCB.

Schematic Design

The Pure Relay Actuators board has four parts: the 12V DC power supply (the usual power supply of the used actuators), the 3.3V DC power supply, the relay management and the Microcontroller.

This board is directly connected to the 230V AC electric grid. The Power Supply unit is based on a 12V Transformer and supplies two power lines: a 3.3V (100mA) line to supply the microcontroller and a 12V (300mA) line to supply the actuators.

The 12V DC power supply is composed by a 12V transformer, to reduce the voltage and isolate the electric grid from the rest of the circuit, a rectifier bridge to transform the alternating current into a continuous one, a filter stage to eliminate all the current alternating components, and finally a regulation stage to improve the stability and output impedance of the power supply output signal. This signal is routed to an 8 pin clamp connector where the sensors are connected to obtain the supply they need. A fuse protects the board against peaks of current.

The 3.3V DC power supply is done by a regulator that, from 12V DC, provides a stable 3.3V DC signal to supply the microcontroller.

The relay control circuit is composed of two independent relay outputs, which allow the switching of two different loads.

Each relay is activated by the microcontroller, and each relay output is configurable through a load selector, which allows the load to be switched in 3 different ways:

- Open circuit/12V Power supply: for loads that must be supplied with 12V voltage.
- Open circuit/230V Power supply: for loads that must be supplied with 12V voltage.
- Open circuit/Short circuit: for loads that only need to short-circuit their pins.

The output configuration is selectable by moving two jumpers for each output, in the relative positions 12V, 230V, ACT.

3.2 Specific PCB

This section describes the specific PCB designs of PERSONA project, not included in the previous section.

3.2.1 Switch On/Off Any Device Board

This board is an actuator by itself. It switches on/off any device that operates at 230V AC, 12V DC or has a relay input to be controlled (like home boilers).

3.2.2 Light Actuator and Detector

This board as an sensor and actuator. It detects the state of the light, switch on/off the light and change light intensity (dimmer).

This board is the responsible for managing the light of the home according to the user preferences. The Light Actuator board can be integrated at the wall, inside the electric boxes, or at the floor thanks to the plastic box where is placed.

References

[1] Fides-Valero, Á., Freddi, M., Furfari, F., Tazari, M.-R.: The PERSONA Framework for Supporting Context-Awareness in Open Distributed Systems. In: Aarts, E., Crowley, J.L., de Ruyter, B., Gerhäuser, H., Pflaum, A., Schmidt, J., Wichert, R. (eds.) AmI 2008. LNCS, vol. 5355, pp. 91–108. Springer, Heidelberg (2008)

[2] AMIGO EU Project, http://www.amigo-project.org

[3] ASK-IT EU Project, http://www.ask-it.org

[4] SOPRANO EU Project, http://www.soprano-ip.org

Context Aware Management Plateform to Invoke Remote or Local e Learning Services: Application to Navigation and Fishing Simulator

Valérie Monfort and Fayssal Felhi

Abstract. Many companies aim to use Web services to integrate heterogeneous and /or remote application in SOA (service oriented architecture) contexts. The SaaS (Software as a Service) economical model allows to link service consumption and pricing. We aim to consider e learning as a set of services, hosted according to Cloud Computing techniques. Moreover, we are convinced some local or remote services may be invoked according to context and training orchestration may be dynamically adapted on run time. We developped a concrete industrial product development based on Web services and WComp plate form for adaptability.

Keywords: Context awareness, SOA, SAAS, Cloud Computing, Web services, e learning.

1 Introduction

Economical context impacts companies and their information system (IS). Companies take over other competitors or develop new business skills, delocalize whole or part of their organization and their production. IS are faced to these genuine constraints and have to overcame these changes. Service oriented architecture (SOA)

Valérie Monfort
MIRACL, Université de Sfax, Tunisie

Valérie Monfort
Université Paris 1 - Panthéon – Sorbonne
e-mail: valerie.Monfort@univ-paris1.fr

Fayssal Felhi
ISIG Kairouan, Khemais Elouini Road, Tunisie
e-mail: felhi_fayssal@yahoo.fr

J.C. Augusto et al. (Eds.): ISAmI 2010, AISC 72, pp. 157–165.
springerlink.com

[17] offers a great flexibility to IS because each application owns interfaces masking implementation details. So, applications own interfaces including services and are seen as black boxes independently connected to a middleware as enterprise application integration bus (EAI) with its connectors and adaptors. However, this integration solution does not allow connecting heterogeneous applications or infrastructures, as distant IS. It is the reason why Web services are based on standards and are till now the cheapest and simplest solution to support interoperability between platforms. Based on Web services, enterprise services bus (ESB) [18] is a kind of services Web based EAI and allows loosely coupling with low costs.

Faced to these technologies, the SaaS (Software as a Service) model allows associating a price to a service consumption. Moreover, Cloud Computing, very linked to SOA and SaaS paradigms, offers delivering IT resources "on demand" by using virtualization paradigm. We are convinced SaaS and Cloud Computing are the fitted solutions to be used for e learning domain where any users as: companies, schools,... could pay services to train learners via Web services and ESB. Therefore, we are developing a new kind of e learning platform, based on orchestrated services, with services used "On Demand", according to training needs and learners skills. This solution aims to address: i) Full services oriented solution to provide full interoperability between authoring tools based on specific e learning standards and any kind of execution platforms. ii) Services consumption according to SaaS model and Cloud computing approach, iii) An execution platforms simplification. Platforms become middleware with orchestration engine for distant, remote, simple or complex services hosted locally. Moreover, some services may be invoked according to context (as weather, date, time, user profile, ...) and/or specific events as mobility [23]. Specific context aware plate forms perform events to adapt system behavior. We used WComp [25][27] plate form as a middleware for context adaptation. The background of our research work is industry and users are learners employed by navigation and fishing companies, to train employees on simulators. We shall process as followed. The second section shows the case study. The third section explains Web services, SaaS and cloud computing. The fourth section addresses the adaptability. The fourth section shows related works.

2 Case Study

We aimed to design and implement a genuine fishing simulator. This software was intended to navigation and fishing schools or fishing ship owner companies. Following partial UML model (Fig.1) shows the different services proposed by the navigation and fishing e learning system. This e learning platform provides fisherman diploma for students (learners) in fishing schools and certificate for companies. The learner can subscribe, have a course, and get a diploma after an exam. The Teacher may be human or not. The system can take decision and can send specific events to complicate lesson if the student has a good level. Fig. 2 shows a process where the Teacher is preparing training and is sending it to the student who identifies himself by invoking identification and authentification services, linked to rules manager. While training learner may receive specific events during

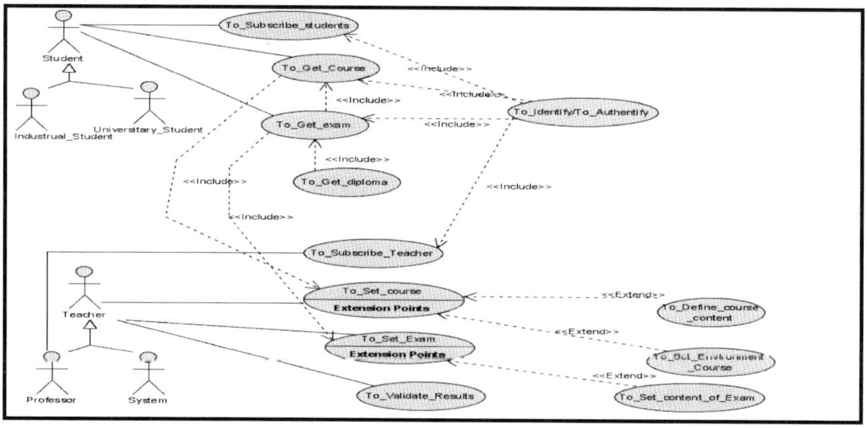

Fig. 1 E learning Services modeling

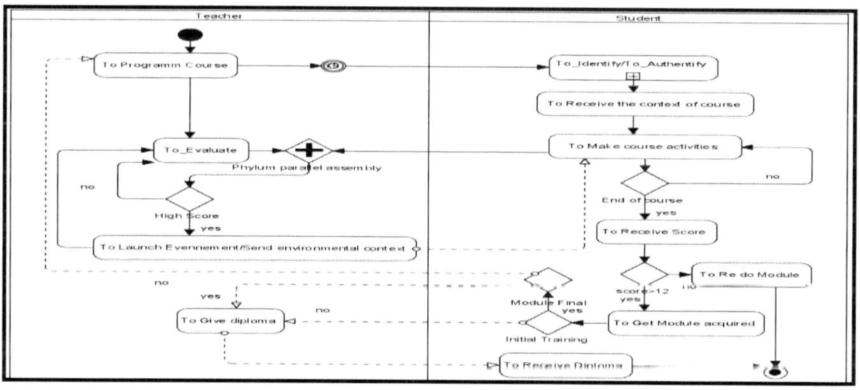

Fig. 2 Training services according to BPMN

current lesson to assess learner's skills. The learner is evaluated at any time. The following e learning architecture (Fig.3) shows services proposed by providers. Actors as learner and teacher can work anywhere and use different media. Services are available according to SaaS model and managed by providers according to Cloud Computing principles as explained in section 3. Contents are defined by the teacher with authoring tools according to standards (see previous section). He defines e learning tasks sequences, and so, e learning services orchestrations. He informs the services providers by loading training content and the concerned learner. The learner may use genuine navigation equipments and/or simulation. The genuine navigation equipments are linked to a middleware able to interpret signals coming from equipments and to send them to Execution platform. The learner may also use PDA or mobile phone. Learner receives the training scenario and while training sends information and invokes services. The e learning

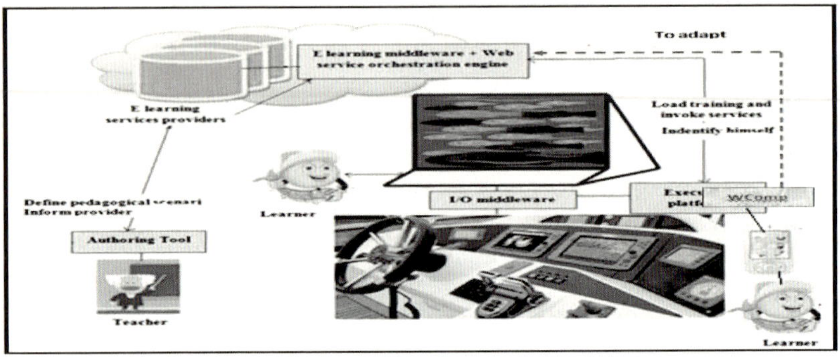

Fig. 3 Introducing WComp

middleware (ESB) manages: routing messages, transporting messages and transforming exchanged data.

The providers repositories manage at least following services as: business, training supervisor, course virtual management, planning management, collaborative management, subscribing management, time/tracking management.

3 Web Services

3.1 Web Services Definition

Web services (WS) [7], like any other middleware technologies, aim to provide mechanisms to bridge heterogeneous platforms, allowing data to flow across various programs[10][11][12]. The WS technology looks very similar to what most middleware technologies looks like. Consequently, each WS has an Interface Definition Language, namely WSDL (Web Service Description Language) [14], that is responsible for the message payload, itself described with the equally famous protocol SOAP (Object Access Protocol)[9], while data structures are explained by XML (eXtended Markup Language) [15]. Very often, WS are stored in UDDI (Universal Description Discovery and Integration) registry [13]. Web services standards are gathered in WSA (Web Service Architecture)[7][8]. WS-BPEL [16] provides a language for the specification of Executable and Abstract business processes. Many integration solutions are based on an ESB. It has services repositories and services orchestration (based on BPMN [19] as Business Process Modeling Notation) that are modeled in a specific modeling tool linked to the environment. Providers use this technology to make increasing business by using SaaS model. Let us see Saas model in following section.

3.2 The SaaS Model

Software as a Service (SaaS) [1] is a model of software deployment whereby a provider licenses an application to customers for use as a service on demand. SaaS software vendors may host the application on their own web servers or download the application to the consumer device, disabling it after use or after the "on-demand" contract expires. The "on-demand" function may be handled internally to share licenses within a firm or by a third-party Application Service Provider (ASP) sharing licenses between firms. "On-demand" licensing and use alleviates the customer's burden of equipping a device with every conceivable application. It also reduces traditional End User License Agreement (EULA) software maintenance, ongoing operation patches, and patch support complexity in an organization.

"On-demand" licensing enables software to become a variable expense, rather than a fixed cost at the time of purchase. It also enables licensing only the amount of software needed versus traditional licenses per device. SaaS also enables the buyer to share licenses across their organization and between organizations, to reduce the cost of acquiring EULAs for every device in their firm.

3.3 Cloud Computing

Cloud computing [8][20][21] is a model for enabling convenient, "on-demand" network access to a shared pool of configurable computing resources (e.g., networks, servers, storage, applications, and services) that can be rapidly provisioned and released with minimal management effort or service provider interaction. This cloud model promotes availability and is composed of five essential characteristics, three service models, and four deployment models as: On-demand self-service, Ubiquitous network access, Resource pooling, Location independence, Homogeneity, Rapid elasticity, Measured service. Cloud computing brings a new level of efficiency and economy to delivering IT resources on demand. It offers efficiency and agility.

Cloud computing is typically divided into three levels of service offerings (fig. 5): Software as a Service (SaaS), Platform as a Service (PaaS), and Infrastructure as a service (IaaS). The middle layer, or PaaS, is the encapsulation of a development environment abstraction and the packaging of a payload of services. PaaS offerings can provide for every phase of software development and testing, or they can be specialized around a particular area, such as content management. IaaS is at the lowest layer and is a means of delivering basic storage and compute capabilities as standardized services over the network. Servers, storage systems, switches, routers, and other systems are pooled (through virtualization technology, for example) to handle specific types of workloads from batch processing to server/storage augmentation during peak loads.

4 Context Adaptation

4.1 The WComp Plate Form

WComp is a prototyping "development" environment for context-aware applications. The WComp Architecture is organized around Containers and Designers paradigms. The purpose of the ***Containers*** is to take into account system services required by ***Components*** of an assembly during runtime: instantiation, destruction of software ***Components*** and Connectors. The purpose of the ***Designers*** allows configuring assemblies of through ***Containers***. To promote adaptation to context WComp uses Aspect [26] Assembly paradigm. Aspect Assemblies can either be selected by a user or fired by a context adaptation process. It uses a weaver that allows adding and or suppressing components. A container includes a set of (Beans) components and each bean has: properties, input methods that use received input information, and output Methods to send to another bean, for instance, output information. Aspect Assemblies allow defining links between Beans by using input and output information. WComp uses UPnP (Plug and Play) technology to detect locally whether the device is active or not and to define input methods and sent events for each component. With this architecture WComp allows: i) managing devices heterogeneity and dynamic discovering by using UPnP, ii) events driven interactions with devices, iii) managing dynamic devices connection and disconnection (dynamic re configuration on run time) in infrastructure. Let us see now the proposed solution.

4.2 Reviewed Architecture

Following architecture shown in Fig. 3, expresses WComp plate form can be imbedded in the execution plate form. WComp intercepts all the events coming from I/O middleware (learner interactions with the equipments) and learner interactions with different media as keyboard and mobile phone. According to these contextual parameters it invokes remote services and new services orchestrations. Let us see the implementation step.

4.3 Implementation

Each Bean owns properties, methods and events that define its status, its role, and its Input/output data (Get/Set). In our example we show the component Bean Boat. Its input method: Valaction expresses a data showing the action to do or the status of moorings or visibility. This component sends a displayed event showing the received data.

Concerning code generated from previous modeling, The lines 46 and 47 of the Fig. 4 show the assembly definition between beans "Boat" and status change (by new events called "eventHandler") in the container. "EventHandler" gets the value

```
46 this.checkBox1.CheckedChanged += new
47 System.EventHandler(this.__checkBox1_to_boat_0);
48 this.boat.EmitStringValue += new
49 Bateau.Bateau.StringValueEventHandler(this.__boat_to_label1_1);
50 [.....]
51 }
```

Fig. 4 WComp implementation for breaking mooring, assembly part

of the “checkedChanged” event that is emitted by “checkbox1” to the input method of the Bean “Boat” called “this_checkBox1_to_boat_0 and that is the parameter of the “EventHandler” method. The lines 48 to 49 are the definition of the assembly between the components beans “Boat” and “label1” in the container. Assembly part allows to re configure components according to training context.

5 Related Works

Several e learning research works [4][5][6] use Web services to get interoperability. [22][24][26] proposed a prototyped architecture with WComp without any access to remote services based infrastructure. No one use SaaS and Cloud paradigms that are the genuine architecture aims of Web services for the industry.

6 Conclusion

We proposed Web services based e learning architecture and we used SaaS and Cloud Computing paradigms to implement a navigation and fishing simulator. The main advantages of this architecture are : i) The interoperability of the platforms and the applications. Applications are considered as black boxes offering services without considering the way they are coded and on which platform they are. Interoperability is promoted by Web services XML based standards. ii) To take into account new needs in terms of new lessons, and to add or modify lessons (flexibility). iii) Lessons and exercises may be accessible via Intranet, Extranet, Internet, also with mobility via different media as lab top, PDA, mobile phone, ... iv) This e learning Architecture is fully distributed, each knowledge component may be supported by one server independently. Based on previous research works, we aim now to propose technical architecture and solutions for context awareness as weather changes during training.

References

[1] The Benefits of SaaS, Web site retrieved from, http://www.computerworld.com/action/article.do?command=viewArticleBasic&articleId=107276

[2] Web site retrieved from, http://ltsc.ieee.org/wg12/files/LOM_1484_12_1_v1_Final_Draft.pdf
[3] Vossen, G., Westerkamp, P.: E Learning as a Web Service, ideas. In: Seventh International Database Engineering and Applications Symposium (IDEAS 2003), p. 242 (2003)
[4] Chepegin, V., Aroyo, L., De Bra, P., Houben, G.J.: CHIME: Service-oriented Framework for Adaptive Web-based Systems. In: De Bra, P. (ed.) Proceedings of Dutch national conference InfWet, Eindhoven, the Netherlands, November 20 (2003)
[5] Ullrich, C., Borau, K., Luo, H., Tan, X., Shen, L., Shen, R.: Why Web 2.0 is Good for Learning and for Research: Principles and Prototypes. In: Proceedings of the 17th International World Wide Web Conference, ACM, New York (2008)
[6] Lu, T., Ullrich, C., Grabowski, B.: Course Generation as a Web-Service for e Learning Systems. In: The 2nd European Young Researchers Workshop on Service Oriented Computing, Leicester, UK, pp. 63–69. University of Leicester (2007)
[7] Web site retrieved from WSA, http://www.w3.org/TR/ws-arch/
[8] Web site retrieved from, http://javaboutique.internet.com/articles/WSApplications/
[9] Web site retrieved from, http://www.w3.org/TR/SOAP
[10] Web site retrieved from, http://www.service-architecture.com/
[11] Web site retrieved from, http://www-106.ibm.com/developerworks/lib-rary/ws-polfram/
[12] Web site retrieved from, http://msdn.microsoft.com/webservices/build-ing/wse/
[13] Web site retrieved from, http://www.uddi.org/pubs/uddi_v3.htm
[14] Web site retrieved from, http://www.w3.org/TR/wsdl20/
[15] Web site retrieved from, http://www.w3.org/XML/
[16] Web site retrieved from, http://www.oasis-open.org/committees/tc_home.php?wg_abbrev=wsbpel
[17] Curbera, F., Khalaf, R., Mukhi, N.: Quality of Service in SOA Environments. An Over-view and Research Agenda (Quality of Service in SOA-Umgebungen). IT - Information Technology 50(2), 99–107 (2008)
[18] Chappell, D.: Enterprise Service Bus. O'Reilly, Alexander Ryan (2006)
[19] Web site retrieved from, http://www.bpmn.org/
[20] Miller, M.: Cloud Computing: Web-Based Applications That Change the Way You Work and Collaborate Online. QUE Publisher (2008)
[21] Reese, G.: Cloud Application Architectures: Building Applications and Infrastructure in the Cloud. O'Reilly Publisher, Sebastopol (2009)
[22] Tigli, J.Y., Lavirotte, S.: Mobility in e learning context: Distributing Information Depending on Geographical Localization and User Profile. In: Proceeding of the 21th ICDE World Conference on Open Learning and distant Education, Hong-Kong, February 18-21 (2004)
[23] Menzel, K.: Mobile Devices in the classroom- Potentials and Requirements. In: ITC@EDU 1st International Workshop on Construction Information Technology in Education, Portoroz, Slovenia (2002)

[24] Stromboni, J.P., Tigli, J.-Y., Sander, P.: Some Experiments for Computer Aided Teaching. Poster in the 20 ICDE World Conference on Open, Dusseldorf, Germany, April 01-05 (2001)

[25] Tigli, J.-Y., Lavirotte, S., Rey, G., Hourdin, V., Riveill, M.: Lightweight Service Oriented Architecture for Pervasive Computing. IJCSI International Journal of Computer Science Issues 4(1) (September 2009) ISSN (Online): 1694-0784, ISSN (Print): 1694-0814

[26] Kiczales, G., Lamping, J., Maeda, C., Lopes, C.: Aspect-oriented programming. In: Aksit, M., Matsuoka, S. (eds.) ECOOP 1997. LNCS, vol. 1241, pp. 220–242. Springer, Heidelberg (1997)

[27] Web site retrieved from, `http://www.oasis-open.org/committees/tc_home.php?wg_abbrev=wsfed`

EVATAR – A Prototyping Middleware Embodying Virtual Agents to Autonomous Robots

Nikolaos Dipsis and Kostas Stathis

Abstract. We present EVATAR, a middleware framework for rapid prototyping of augmented reality, ubiquitous robotic applications. More specifically, EVATAR aims to connect low cost robotic applications running in a physical environment with software agents that implement the functionality of robots in a MAS (Multi Agent System) electronic environment, as if the robots were the avatars of the agents. We exemplify our ideas by showing how EVATAR can enable agents in the GOLEM agent platform to be connected and control Lego robots in order to create a distributed, autonomous robotic system.

Keywords: middleware, prototyping, MAS, robots, avatars, ubiquitous robot.

1 Introduction

Creating Ambient Intelligence (AmI) applications involving multiple robots manifesting complex collaborative and cognitive behaviors to assist humans in everyday life is a challenging task. A common approach to achieve this, is to integrate hardware and software components through such recent robots middleware and development frameworks as [9][3,4][1] and [5]. These frameworks predominantly distribute the computation and system resources among interconnected devices for integrating heterogeneous components while focusing in defining clear well-documented interfaces permitting the reuse of existing code. Although they generally reduce the difficulty and cost associated with prototyping such robotic applications they are typically constraint by the limited computational capabilities of robotic hardware and operating systems.

Nikolaos Dipsis · Kostas Stathis
Department of Computer Science, Royal Holloway, University of London, UK
e-mail: {nikolaos,kostas}@cs.rhul.ac.uk

J.C. Augusto et al. (Eds.): ISAmI 2010, AISC 72, pp. 167–175.
springerlink.com

We are interested in liberating a robotic application from the computational constraints by wirelessly connecting the behavior needed by a robot in an application with the reasoning and decision making of software agents running on a more computationally powerful MAS platform. To interact with humans, these agents commonly feature user friendly, autonomous, intelligent and cognitive behaviors. The above characteristics in the design of such virtual agents coincide with several requirements of AmI systems that employ devices that interact with humans. Consequently, creating ubiquitous agents/robots by integrating virtual software agents with physical robots offers great potential for application to AmI.

In this paper we introduce the EVATAR (Embodying Virtual Agents To Autonomous Robots) middleware, a framework specializing in rapid prototyping of distributed augmented reality robotic applications by embodying virtual software agents to low-cost robots. EVATAR requires from a virtual agent the following: a) a component acting as a virtual body and situating the agent in an electronic environment so that it can interact with the environment and other agents via virtual sensors and actuators b) a component acting as a "mind" with an internal state of the agent for controlling the agent's body behaviors. On the other hand, robots should be mobile, have sensing capabilities and utilize a remote control API (application interface). Based on these requirements, EVATAR permits one-to-one relationships between virtual software agent bodies residing in a MAS with physical robots for remote control, remote sensing and remote manifestation of virtual agent behaviors.

The next section presents EVATAR and section 3 describes an experiment. Section 4 compares EVATAR to existing approaches while Section 5 concludes presenting our future plans.

2 EVATAR

EVATAR is a middleware for embodying software agents into robots. We developed BIL (Body Integration Language), a language allowing agent-robot communication and integration. BIL is used within EVATAR to provide a common language by translating requests between agent and robot bodies. It is designed to facilitate the manifestation of physical presence for virtual agents by extending their virtual sensors and actuators to the physical ones of robots.

The EVATAR architecture features three main elements: the *Remote Entity*, the *Remote Mediator* and the *Interconnection Layer*. An agent or robot qualifies as a *Remote Entity* if it uses an *Interface* (EVATAR API) to EVATAR. Architecturally, EVATAR is implementing an adaptation of the Mediator Pattern [10] defined as a behavioral design pattern allowing the mediation of networked remote entities(see Fig. 2) featuring loose coupling and ensuring interoperability and scalability.

Despite the fact that the middleware follows the centralized Mediator approach for the exchange of signals, the actual applications that utilize EVATAR are of distributed nature. In a typical EVATAR application the software agents, the robots and the middleware component itself are treated as distributed processes in a wireless network interacting with each other via the EVATAR middleware. EVATAR coordinates the exchange of signals among the distributed entities but it does not

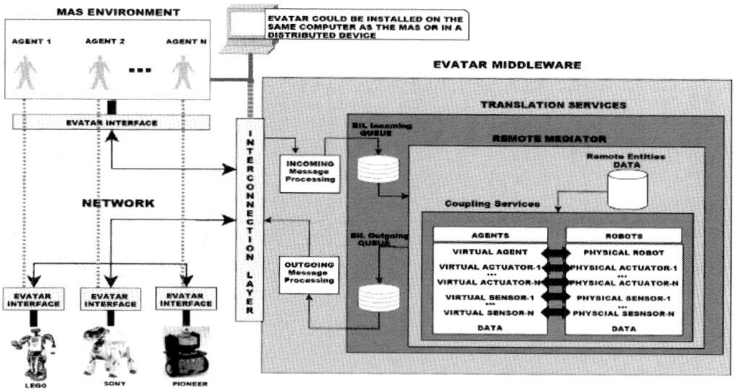

Fig. 1 EVATAR Overview

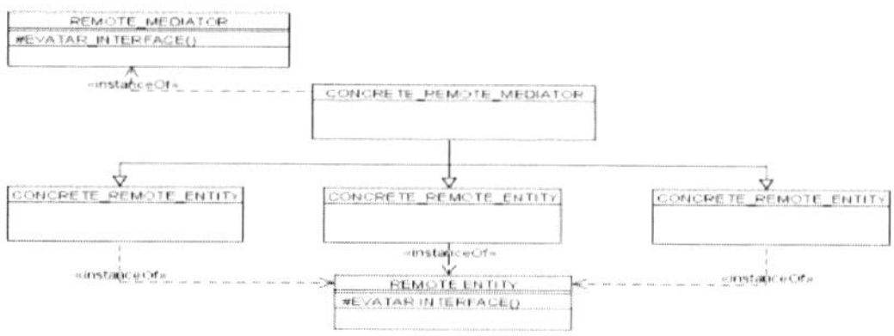

Fig. 2 EVATAR Architecture

control the application. The control is assumed by these decentralized entities. To describe protocols for the exchange of signals between the bodies of software agents and robots we use BIL (Body Interaction Language). The structure of BIL signals is shown in Table 1:

Table 1 The elements of a BIL signal

SenderID	The identifier of the sender of the signal.
ReceiverID	The identifier of the receiver of the signal.
Context	The context of the communication e.g. request awaiting feedback, feedback, request not awaiting feedback, data passing signal etc.
ContentType	The type of the signal content denoting: actuator integration, sensor integration, sensor-actuator integration, Data passing.
Content	The content of the signal, typically an action description and its associated parameters for remote acting and sensing, including the Body Part Identifiers.

Signals are described in XML. In Fig. 3 we can see an example of a BIL signal relaying the message from a virtual agent to a Lego Robot for setting the rotation speed of a particular motor to value 4 for 3 seconds.

```
<?xml version = "1.0" ?>
<EVATAR>
<SenderID> Agent002 </SenderID>
<ReceiverID> Robot001 </ReceiverID>
<Context> AgentRequestWithFeedback </Context>
<ContentType> ACTUATOR </ContentType>
<Content>
   <SenderBodyPartID> VirtualActuator01 </SenderBodyPartID>
   <ReceiverBodyPartID> RobotActuator01</ReceiverBodyPartID>
   <RotationSpeed> 4 </RotationSpeed>
   <Duration> 3 </Duration>
   <Angle> -1 </Angle>
   <Speed> -1 </Speed>
</Content>
</EVATAR>
```

Fig. 3 BIL signal example for Lego motor

Agents and robots need to include the EVATAR API to participate in an application and gain access to functions that construct or handle BIL signals. Through this API the system developer defines: a) identifiers for each participating agent and robot as well as for each of their corresponding virtual and robotic sensors and actuators and b) one-to-one exclusive mappings of the identifiers of virtual agents, sensors and actuators to the identifiers of robots, robotic sensors and actuators (Body Mapping, Fig. 4). The BIL structure is key to understanding the manifestation of the three elements of embodiment in EVATAR: *Binding*, *Acting/Sensing* and *Sensory Input.* In a typical embodiment lifecycle (Fig. 5), EVATAR realizes the Body Mappings by matching the above identifiers contained in BIL signals as the SenderID, ReceiverID, SenderBodyPartID and ReceiverBodyPartID fields of Table 1. Thus, EVATAR is able to *bind* virtual agents and body parts with robots and robotic components in order to route and coordinate the exchange of BIL signals. The communication during the embodiment life-cycle follows the loose coupling featured by the Mediator.

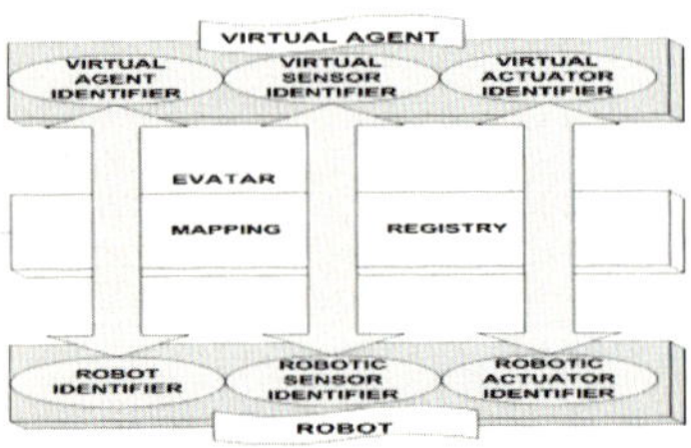

Fig. 4 Body Mapping in EVATAR

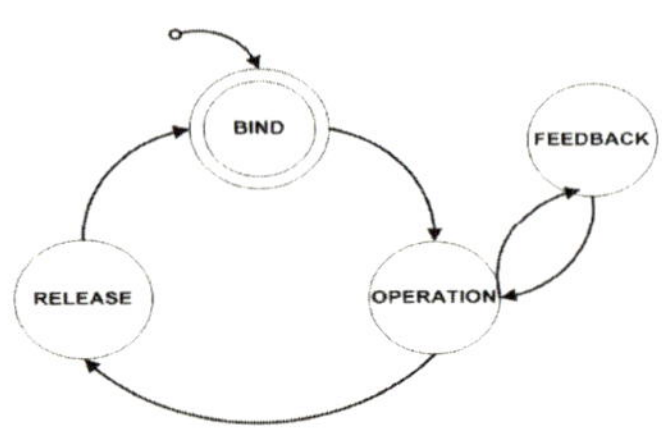

Fig. 5 Embodiment Life-Cycle

Binding ensures that the correct robot will receive a BIL signal. The "Content" data of the signal holds the "body part" identifiers (see Content Table 1) for binding the correct physical sensor or actuator to the virtual one of the sending agent (Fig. 4) along with parameters for the interface on the remote control API of the robot to control this physical sensor or actuator for a remote *operation* (Acting/Sensing).

When a computer gamer manipulates an Avatar, the sensing of the Avatars virtual environment is usually derived by visual effects on the screen. EVATAR allows the virtual agent perception of a physical environment by replacing the virtual sensory input with the one deriving from the physical sensor of the robot (Fig. 6) using BIL *Sensory Input* signals (*feedback*, Fig. 5). If a relationship is *released* the robots return to a waiting for a new embodiment cycle state.

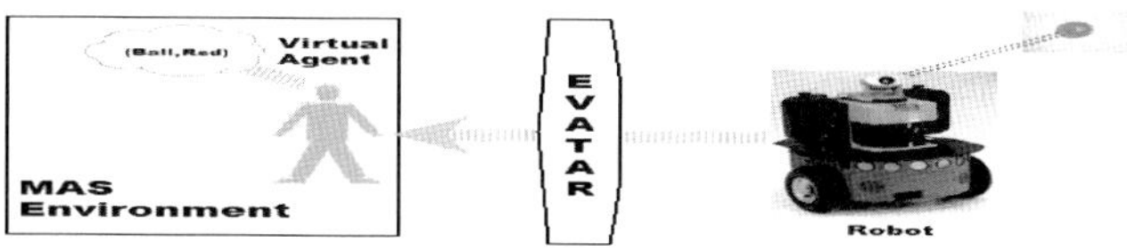

Fig. 6 An agent sensing the color of a ball through EVATAR

The Network Interconnection Component in the EVATAR API and the main EVATAR functionality uses a two-way RMI (Remote Method Invocation) [8] implementation to achieve connectivity among the distributed entities within the network. While the MAS and the robots do not "talk" directly, the Interconnection Layer features dual RMI instances between each of them and the Remote Mediator. Currently EVATAR supports JAVA based MAS and robots due to its RMI implementation. More details on the interconnection are beyond the scope of this work.

3 Experimentation

To demonstrate the use of EVATAR we experiment with linking 2 Lego robots [6] and 2 agents in the GOLEM MAS platform [7]. The Lego Robots included in their body three actuators, a light sensor and a touch sensor. The software programs running on the Lego robots were the custom Lego firmware and the JAVA-compatible remote control API (LEJOS) that features remote control of the Robots motors and sensors. GOLEM is a MAS environment for situating cognitive agents. The agent interaction is mediated by the environment, which is an evolving composite structure supporting Agent-Agent and Agent-Object interaction. The agents in GOLEM feature two main components (a) a "mind" component giving the agent reasoning and decision-making capabilities and (b) a "virtual body" containing virtual sensors and actuators for interaction with objects in the virtual environment.

The GOLEM agents were programmed to include virtual actuators for movement and "grabbing" of objects, a virtual sensor for acquiring color values of virtual objects and a logic-based mind for planning the agents actions based on their communication and reasoning capabilities. The agents internalized unknown objects by storing their perceptual descriptions in order to be able to identify and communicate regarding them. We included the EVATAR interface on GOLEM and the Lego API software (running in different machines) while the EVATAR middleware was installed on the same computer as GOLEM. After installing the API we defined identifiers for every agent, Lego robot and their corresponding virtual and physical sensors and actuators by calling the appropriate interface function and we registered the exclusive mappings between them.

The experiment's main goal was to enable the manifestation of the agent behaviors by the Lego robots. EVATAR is designed to follow the evolution in the state of the MAS. For example, when a GOLEM agent decides to change its position, in order to instigate the change in the position of the robot we call the interface methods of EVATAR in the MAS code to create BIL signals such as the one in Fig. 3. The BIL signal contains the identifiers needed by EVATAR to route this signal from the sending GOLEM agent to the EVATAR Interface of the corresponding Lego robot API. As soon as it receives this signal, the EVATAR interface installed on the Lego API utilizes the "Content" part of the signal to identify which sensor or actuator is bound to the virtual one of the GOLEM agent. It also utilizes the rest of the parameters of the "Content" data in order to invoke a control function on the Lego API. In our example in Fig.3, the signal with sender Agent002 will be received by the robot with receiver Robot001 in order to perform an operation with the actuator with identifier RobotActuator01. The *RotationSpeed* field sets the speed of the Lego motors rotation to 4 and this rotation will last for 3 seconds according to the *Duration* field.

EVATAR augments the GOLEM agent sensors to the corresponding Lego sensors. This is achieved via the two types of BIL signals: a)remote sensing signals from the virtual agent and b) sensory data signals (feedback) from the robot back to the agent. In our application the virtual sensor for acquiring color values of virtual objects of the GOLEM agent is bound to the Lego light sensor. The agents request the robots to sense the area of vicinity by sending a BIL signal. The input of their virtual sensors is thus replaced by the level of luminosity from the Lego sensors sent as a sensory feedback BIL signal by the robot.

The GOLEM-EVATAR-Lego system was situated in an environment with three unrecognized objects. GOLEM included three virtual agents, two featuring Avatars in the physical world as Lego robots. According to the scenario, the agents embodied in the robotic avatars internalized the objects in the system, by storing their perceptual description (color approximation derived by the light sensor) and performed actions upon them. The virtual agent without an avatar chose newly learned objects and instructed others with robot-avatars able to perform actions upon them. Besides the more complicated behaviors of reasoning, planning and communication taking place within the MAS, the following physical world behaviors where manifested in the real world environment by the robot-avatars (Fig. 7): a) directional movement (e.g. forward, left, right, forward-left, backwards etc.) b) detect an object c) pick an object d) combinations of the above for more complex behaviors. In order to achieve the above behaviors in our application we send combinations of BIL signals.

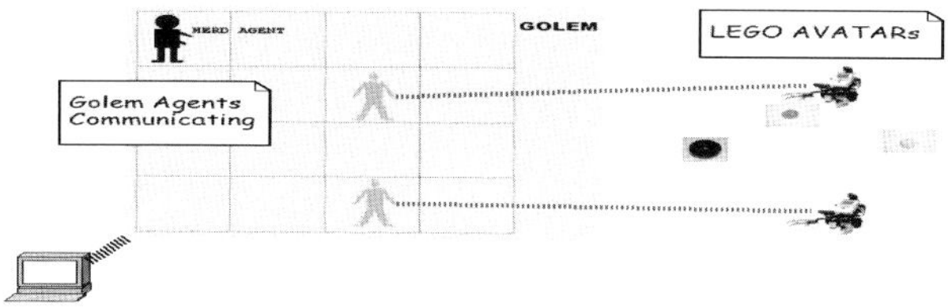

Fig. 7 GOLEM Agents communicate and plan the actions of their Lego Avatars

4 Related Work - Discussion

The experiment provided useful feedback regarding the potential of EVATAR for migrating software MAS to augmented reality systems using physical robots. In the introduction we identified middleware frameworks capable of such applications. MIRO the "Middleware for Robots" [1] enables communication among a robot's different modules, between several robots and between a robot and a distributed device such as a PC. Other frameworks are the Microsoft Robotics Studio[9] and the open source Player Stage Project[3,4], robotic development platforms featuring the simulation of multi-robot behaviors in virtual environments and the ability to reproduce the mechanical behaviors of the robots. MARIE[5] is a robotic middleware following a conceptually similar approach to EVATAR utilizing a higher level of abstraction. It features a framework for integrating new and existing software for rapid prototyping Robotic applications in a distributed environment. It exploits the diversity of communication protocols and mechanisms thus allowing interoperability between different software.

EVATAR does not intend to compete with these middleware for providing better communication and interfacing between more heterogeneous devices, software platforms, sensors and actuators. Instead it specializes in augmenting existing virtual MAS and robots thus incorporating the benefits of utilizing Ubiquitous agents. Also, EVATAR allows the creation of affordable applications of this specific type with minimal effort due to its specialization, while the above frameworks still require the development of custom code for such systems. They may enable for example a MAS to send a message to a Robot but making this message usable and in general creating an application featuring the integration of the Robots and the MAS into a system would still require both development time and effort.

The PEIS ecology[12] provides a framework for integrating autonomous robots in AmI systems. It utilizes the PEIS middleware by which devices of different types and capabilities (including robots) can cooperate using a distributed tuple-space. PEIS is purely focused on the integration of the robots with the rest of the devices by simplifying their communication and cooperation and enriching their sensing capabilities in a tagged environment. EVATAR features a different approach, since it attempts to incorporate the benefits of utilizing software agents for integrating autonomous robots in ambient intelligent environments. Therefore, even though they both attempt to integrate robots in intelligent environments, they follow different approaches focusing on different types of problems.

An interesting approach for enabling the communication between software agents and physical robots is the Middle Layer for Incorporations among Ubiquitous robots by Kim, Choi and Lim [11]. It is similar to EVATAR since it enables the remote control of physical robots as avatars of software agents. The middle layer consists of a sensor and a behavior mapper. The first helps software robots obtain physical sensor information from mobile robots while the second allows software robots to present physical behavior. The key difference is that EVATAR provides a framework for the embodiment of virtual sensors and actuators featuring virtual control and sensing of physical avatars whereas the "Middle Layer" in [11] is focused on mapping behavior and sensor names with no control of how this is achieved by the robots.

5 Conclusions – Future Work

EVATAR provides a framework for further research into applications of mixed reality environments where agents or robots may interact with real or virtual objects. We aim to experiment with EVATAR and obtain further results and metrics regarding the effectiveness of EVATAR in comparison to other approaches. Central to our future plans, EVATAR will be enhanced in terms of features and portability in order to support our ongoing research in ubiquitous agent and robotic applications. Finally, inspired by the work of Perez et al [13], we intend to investigate the potential benefits for EVATAR based applications by enhancing BIL signals to instigate more complex behaviors offered by Robots.

References

1. Utz, H., Sablatng, S., Enderle, S., Kraetzschmar, G.: Miro – Middleware for Mobile Robot Applications. IEEE Transactions on Robotics and Automation 18(4), 493–497 (2002)
2. ULM Sparrows, `http://www.sparrows.uni-ulm.de/`
3. Gerkey, B.P., Vaughan, R.T., Howard, A.: The Player/Stage Project: Tools for Multi-Robot and Distributed Sensor Systems. In: Proceedings of the International Conference on Advanced Robotics (ICAR 2003), Coimbra, Portugal, pp. 317–323 (2003)
4. Player STAGE and GAZEBO: `http://playerstage.sourceforge.net/`
5. Côté, C., Létourneau, D., Raïevsky, C., Brosseau, Y., Michaud, F.: Using MARIE for mobile robot software development and integration. In: Brugali, D. (ed.) Software Engineering for Experimental Robotics. Springer Tracts on Advanced Robotics (2006b)
6. Lego Co. Mindstorms Robotics Manual, `http://www.legomindstorms.com`
7. Bromuri, S., Stathis, K.: Situating Cognitive Agents in GOLEM. In: Weyns, D., Brueckner, S.A., Demazeau, Y. (eds.) EEMMAS 2007. LNCS (LNAI), vol. 5049, pp. 115–134. Springer, Heidelberg (2008)
8. RMI: `http://java.sun.com/javase/technologies/core/basic/rmi/index.jsp`

9. Jackson, J.: Microsoft robotics studio: A technical introduction. IEEE Robotics & Automation Magazine 14, 82–87 (2007)
10. Gamma, E., Helm, R., Johnson, R., Vlissides, J.: Design Patterns: Elements of Reusable Object-Oriented Software. Addison-Wesley, Reading (1994)
11. Kim, T., Choi, S., Lim, J.: Incorporation of a Software Robot and a Mobile Robot Using a Middle Layer. IEEE Transactions on Systems, Man, and Cybernetics, Part C: Applications and Reviews 37(6), 1342–1348 (2007)
12. Broxvall, M., Gritti, M., Saffiotti, A., Seo, B.S., Cho, Y.J.: PEIS Ecology: Integrating Robots into Smart Environments. In: Proc. of the IEEE Int. Conf. on Robotics and Automation (ICRA), Orlando, USA (2006)
13. Perez, J., Ali, N., Carsi, J.A., Ramos, I., Alvarez, B., Sanchez, P., Pastor, J.A.: Integrating aspects in software architectures: PRISMA applied to robotic tele-operated systems. Information Software Technology, 969–990 (2008)

User Recognition in AAL Environments

Ricardo Costa, Paulo Novais, Ângelo Costa, Luís Lima, and José Neves

Abstract. Healthcare projects that intend to decrease the economical and social costs of the real ageing population phenomenon, through the de-localisation of healthcare services delivery and management to the home, have been arising in the scientific community. The VirtualECare project is one of those, so called, Ambient Assisted Living environments, which we have taken a step forward with the introduction of proactive techniques for better adapting to its users, namely elderly or chronic patients, once it is able to learn with their interaction based in contexts. This learning, however, causes the system need to know with whom it is interacting, especially if we are dealing with multi-user environments. Basic detection techniques based in possible devices that users carries along with them (e.g. RFID tags, mobile phones, ...) are not good enough, since they can lose/forgot/switch them. To obtain the expected results the technology used has to be more advanced and available in several platforms. One possible and already fairly developed technique is Facial Recognition, and it appears to be the most appropriate one to handle the problem.This document exposes the initial approach of the VirtualECare project to the Facial Recognition area.

Keywords: Ambient Assisted Living, e-Health, Face Recognition.

Paulo Novais · Ângelo Costa · José Neves
CCTC, Departamento de Informática, Universidade do Minho, Braga - Portugal
e-mail: pjon@di.uminho.pt, acosta@di.uminho.pt, jneves@di.uminho.pt

Ricardo Costa · Luís Lima
CIICESI, Escola Superior de Tecnologia e Gestão de Felgueiras, Instituto Politécnico do Porto, Felgueiras - Portugal
e-mail: rcosta@estgf.ipp.pt, llima@estgf.ipp.pt

J.C. Augusto et al. (Eds.): ISAmI 2010, AISC 72, pp. 177–184.
springerlink.com

1 Introduction

The population ageing phenomenon [1] and specially its consequences in the health-care and social security systems around the world, particularly in developed countries, associated with the need in providing elderly with a quality and dignified active ageing solution [2] constitutes a real and urgent problem in today's society. The rapidly ageing societies claim for better ways to treat the elderly persons with more dignity and consideration. As the ageing tendency keeps climbing, the mild-age persons are already preparing their future towards a most peaceful and supported way. This urgency raised the need for a rapid response from the several healthcare sectors, direct or indirect, players that, in most cases, have turned to the IT sector searching for answers. In this scenario the VirtualECare project [3] appeared as one possible solution, in search of an active ageing that can cope with the demands of the future of the soon-to-come elderly persons.

1.1 Active Ageing

According to the World Health Organisation (WHO) [2] today's society must introduce new strategies for the population ageing, as elderly population is an asset that the society is starting to properly considering, in particularly the senior professionals witch possess a number of skills and accumulated experience that need to be transferred to younger ones and/or participate in social activities [4]. In developed countries the core of the community that has 60 years old and upwards is constituted already by well educated people that has many contact with new technologies and is used to use modern devices and services, things that are already a necessity in their everyday life. Additionally, recent studies [1] show that by keeping the human been active and engaged with activities of low and high complexity is beneficial not only to the body as also, and perhaps more importantly, to the mind, so keeping an active life helps the person in several levels and fields of health.

1.2 VirtualECare

The VirtualECare project (Figure 1) envisions a new and effective way of providing healthcare, where the treatment will no more be institution centred but, instead, will shift to be user centred, resulting in a better and cheaper service for both the user and the provider. Although at first glance this may seem a contradiction, the lower of costs is possible through the use of new, low-cost, devices and technologies that allow the monitoring and assessment of patient vital data and location in real time. This same data, after adequate processing, can conduct to alarms, recommendations or even actions to assist the needed patient while he maintains is normal, day-to-day, life. These new healthcare provision approximations bring great advantages that may also be extended to relatives and friends of the patient, since they can be informed, in real time, of his actual condition allowing them to also intervene in case of necessity [3, 5].

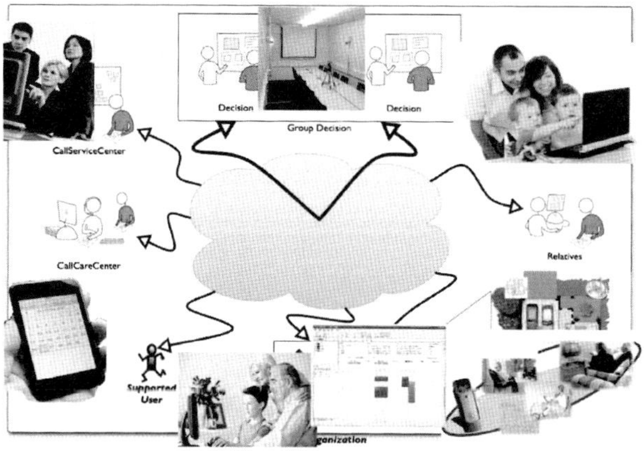

Fig. 1 The VirtualECare Project

As a consequence of this new way of healthcare provision we had the need to enrich the patient home with Ambient Intelligence (AmI) and, more concretely, Ambient Assisted Living (AAL) technologies. This approximation allowed us to better respond to the patient specific needs, modulating his home environment as needed in order to better respond to his expectations [6]. In an early stage of the system development, and because of the initial focus of the conception and targeted community, only was projected the use and activation of a singular person for each house or implementation. However, as the project evolved and new facts and community feedback where received, for instance the deployment of the system in elderly/community centres, health centres and hospitals or the system need maintain a user profile in order to be able to learn with is actions according to the actual context, it was realised that there was a missing part ready to be constructed, now that we have reached a more mature implementation. We started to realise the importance of implement techniques that will allow our system not only to be capable of automatically, and without any user/patient intervention, recognise the presence of different individuals [7] but also to uniquely identify them. This individual recognition, crucial in those multi-user environments, will contribute to the set of different sub-environments that can match to different users and seamlessly work and contribute to the expected results.

An RFID module is already been developed in the VirtualECare project context but its focus tends to another area where the application is more oriented to an office-like environment, in which the persons and devices that are applied are very different to the ones we approach in this presented case. Here, it is not needed any control over moving objects or very frequent log-ins and log-outs. There are several cases that misfit the use of the RFID in this scenario, the most important is the security. The RFID tag can be lost or misused by another person than the user, whilst the persons face cannot be lost or even traded.

2 Recognition

With the increase of the Information and Communication Technologies in all aspects and sectors of our life, a world of unprecedented scenarios is arising. Some of these scenarios are the explosions of new and different techniques of automatic and seamless individual's recognition. These techniques, used in several areas of our society (e.g. airports, industry, law enforcement), can also be applied in AmI and consequently AAL environments, not only to regulate controlled access to them, but also, as a consequence of the possibility to uniquely identify an individual, to make the necessary automatic adaptations to it, according to that specific individual needs (according to his pre configured profile). Nowadays there are several developments already in the domestic area. The personal computers, such as laptops already bring fingerprint sensors and Facial Recognition, although they are fairly primitive in terms of the functioning and detection, mainly because the companies do not use the best hardware available. Facial Detection is now the top area of the Biometric sector. It is practical since no physical interaction has to be made and if the cameras comply with high speed shuttering several faces can be processed in just a few seconds.

2.1 Biometric

Biometric recognition uses our distinctive physiological (e.g., fingerprints, face, retina, iris) and behavioural (e.g., gait, signature) characteristics for automatic recognising. These biometric distinctive physiological and behavioural characteristics cannot be easily misplaced, forged and/or shared [8]. They are, inclusively, considered more reliable for person recognition than any other traditional token or knowledge-based methods making biometrics very suited for AmI/AAL environments, and this is especially true for a biometric physiological characteristic such as face which is one of the most common methods of recognition that humans use in their visual interactions, and allows to recognise the user in a not intrusive way without any physical contact with the sensor [9]. Using several points of recognition the system can construct a valid profile of the possible user, being registered in the main system the information can spread to the remaining system without the interaction of the user, being available at once the capacity of logging into the system.

Fig. 2 Facial Recognition

2.2 *Facial Recognition*

Facial recognition, as above already mentioned, is one of the possible biometric physiological characteristics that can be used to uniquely identify a person. Normally, these kind of systems are on 3D features to verify the identity of subjects accessing the controlled AmI environment and to customise the available services accordingly. In other terms, it allow us to add a social dimension to man-machine communication and thus may help to make such environments more attractive to the human user [9].

The facial signature can be captured nowadays by a simple web-cam or camera as they process enough image quality and speed do do it so. The facial pinpoints consist in the marks and features of the user face, usually nodal points. In terms of the 3D facial recognition, the technology this project intent to embrace, several captures of the face are taken in different angles and facial expressions, mapping a more realistic approach to the user face. This is done so that the user can be more rapidly and accurately detected, being in the most advanced cases possible to recognise the head silhouette and in fairly low light conditions, with the help of infra-red cameras. With the independence of colours and natural light conditions it is safe to say that this type of recognition can work in any normal in-house conditions, not to mention the security in recognition, provided by the accuracy that this system has. After the initial capture and processing any modern camera is able to do the system log-in, being only advised the use of infra-red lights to stabilise the surrounding light conditions and create a mini environment in terms of light exposure.

The enrolment of the several captures or the 3D scan make a perfect model that has millions of localisation points, these points are taken as markers and if the face is presented to the scanner in a awkward position, because there are several referential points the annalist agent can mismatch the received image to a single point and construct the rest of the face by pattern matching, applied by the surface texture analysis model.

3 VirtualECare Recognition Module

The VirtualECare system architecture (Figure 3) was idealised to be a modular one, in order to allow the addition of new modules in the future, in the measure of necessity and system evolving. Heavily based in the OSGi technology and decentralised systems this project has the capacity of receiving new modules, either locally or in the central system (cloud). Implemented under the Multi-Agent paradigm all the modules and agents in the system communicate with each other creating and intricate system of communication and information sharing. Each module can have several agents and control them independently but it keeps connection with the remaining modules in a meta-communication form. Several modules are already working in both system types, being a sensors monitoring system and a user intelligent calendar organiser two of the most relevant modules and actually well beyond the proof-of-concept phase. The project has already the ability to identify a user based in some kind of device he was carrying, like a RFID tag or a mobile phone, however this was not a ideal situation as users tend to forget to carry them around the house and/or makes very difficult the task of knowing their exact location, typically the target users of this project.

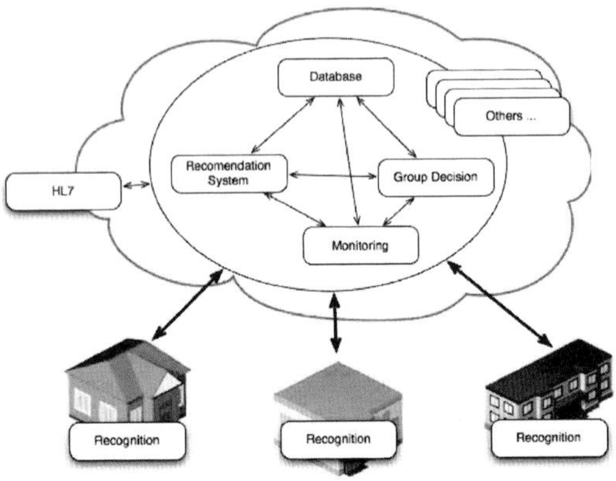

Fig. 3 VirtualECare Modular Architecture with the Recognition Module

3.1 Implementation

Due to the modular characteristic, theoretically, it should be trivial to add any kind of recognition module that meets the project technology standards. However, during the definition and architecture phase of the recognition module, we realised it should not stand in the "cloud" like all the others, once the recognition should take place in each different environment, despite the user identification being needed in the entire system.

Due to this fact, we opted to deploy the recognition modules inside each user environment (Figure 4), interconnected with the existing local technologies (e.g. OSGi and Agents). This decision also allowed us to be able to keep the base characteristics of the environment working (including recognition) even if, for some reason, the connection to the "cloud" was lost. In terms of the technology point of view there can be a camera in each room or terminal or only one to the entire system, this can be adapted to each environment/user necessity. As there will be agents locally controlling the capture of the information the preservation of the information will be secured and then relayed to the "cloud" to be conveniently logged and processed. Each user has to be inserted in the system by a technician and all the persons that sporadically come in touch with terminal can request to the support line an authorisation or use a predefined "guest" log-in.

The availability of this system can be used to control the operation of the several devices present in the house, activated personalised profiles and change the devices

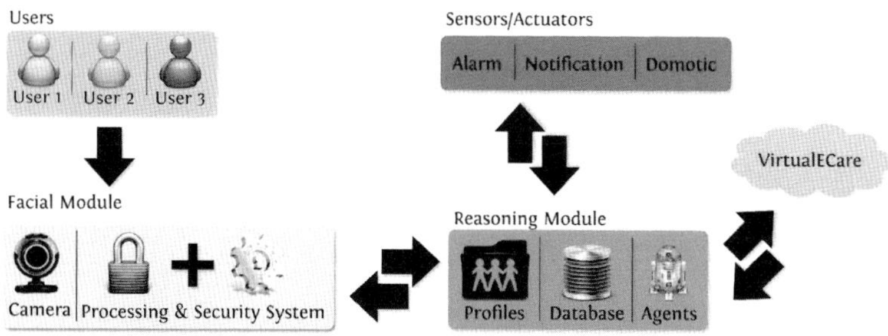

Fig. 4 Designed Architecture

configuration in each room the user enters, open automatically the front door without the need of a key, keep track of where the user enters and exits between other several hypotheses available.

3.2 Security

In terms of the added security value, it is of the most interesting factor. As it was previously mentioned the face is one thins that is hard to forge or manipulate so this system is of the uttermost security level. In the verification phase it is assured that already the user to be logged is in fact the user registered in the system. The use of the cameras can also be used in several other purposes, like monitoring every person that approaches the front door of the house and register them in the database for further analysis and processing, being one of the abilities the easy introduction of that person in the system, if the user find it useful. Furthermore the use of the cameras can also be used in an extreme case to monitor and report to the right service/person of possible cases of urgency or emergency happening the the user home. Any stranger reported to the system can be automatically notified to the user or relatives for authorisation or it refuse.

4 Conclusions and Future Work

In this paper we present the idea of bringing the, already available, new technologies and techniques used for biometric identification, specially the ones based on facial recognition to the VirtualECare project. We believe the introduction of such techniques in AmI/AAL environment will open new horizons and allow such environments, not only to be better accepted by its users, but also to more adequate respond to their needs and expectations. The introduction of the Recognition Module, based in the above presented Facial Recognition Techniques, in the VirtualECare system brought several new facilities and user friendless, thus contributing to one of the objectives of AmI environments. We are now able to uniquely identify each user of the system without his interaction, and even if for now, that may not seem very impressive, the addition of this simple capacity opened our horizons, since we can start a new line of research for multi-user support inside a single environment and, for instance, adequate different rooms environment to the user inside it. Obviously, the future introduction of this new recognition module will increase the solution cost, but we hope to demonstrate the gains associated with maximal user dematerialisation of mobile devices and with the system additional personification.

References

1. Desa, P.D.: World Population Ageing: 1950-2050, United Nations (2001)
2. W. H. Organization, Active Ageing: Towards Age-Friendly Primary Health Care. World Health Organization (2004)
3. Costa, R., Novais, P., Lima, L., Carneiro, D., Samico, D., Oliveira, J., Machado, J., Neves, J.: Virtualecare: Intelligent assisted living. In: Electronic Healthcare, pp. 138–144 (2009)
4. Camarinha-Matos, L., Afsarmanesh, H.: The Need for a Strategic R& D Roadmap for Active Ageing (2009)
5. Costa, R., Novais, P., Lima, L., Cruz, J.B., Neves, J.: Virtualecare: Group support in collaborative networks organizations for digital homecare. In: Handbook of Digital Homecare, pp. 151–178 (2009)
6. Carneiro, D., Costa, R., Novais, P., Neves, J., Machado, J., Neves, J.: Simulating and monitoring ambient assisted living. In: Proceedings of the ESM 2008 - The 22nd annual European Simulation and Modelling Conference (2008)
7. Ducatel, K., Bogdanowicz, M., Scapolo, F., Leijten, J., Burgelman, J.: Scenarios for ambient intelligence in 2010. Office for Official Publications of the European Communities (2001)
8. Maltoni, D., Maio, D., Jain, A., Prabhakar, S.: Handbook of Fingerprint Recognition. Springer, New York (2003)
9. Abate, A., Ricciardi, S., Sabatino, G.: 3d face recognition in a ambient intelligence environment scenario. Face Recognition (2007)

Patient Monitoring under an Ambient Intelligence Setting

Pedro Eduardo Gomes, Alberto Marques, Ângelo Costa,
Paulo Novais, and José Neves

Abstract. In recent years there has been a growing interest in developing Ambient Intelligence based systems in order to create smart environments for user and environmental monitoring. In fact, higher-level monitoring systems with vital information about the user and the environment around him/her represents an improvement of the quality of care provided. In this paper, we propose an architecture that implements a multi-agent user-profile based system for patient monitoring aimed to improve the assistance and health care provided. This system mixes logical based reasoning mechanisms with context-aware technologies. It is also presented a case based on a scenario developed at a major Portuguese healthcare institution.

Keywords: Ambient Intelligence, eHealth, Patient Monitoring, Multi-Agent Systems, Case Based Reasoning, RFID.

1 Ambient Intelligence in Healthcare

The emergence of the computer as a tool for the masses has enabled its generalization. The advent of the microprocessor significantly increased computing power, being now embedded in objects we use on a daily basis, such as mobile phones, PDAs

Pedro Eduardo Gomes · Ângelo Costa · Paulo Novais · José Neves
CCTC, Departamento de Informática, Universidade do Minho, Braga - Portugal
e-mail: p.eduardogomes@gmail.com, acosta@di.uminho.pt,
pjon@di.uminho.pt, jneves@di.uminho.pt

Alberto Marques
Chairman of the Board of Directors - CHTS, Penafiel, Portugal
e-mail: josealbertomarques@gmail.com

J.C. Augusto et al. (Eds.): ISAmI 2010, AISC 72, pp. 185–188.
springerlink.com

or Satellite Navigation systems [1, 2]. This development gave rise to a new area: Ambient Intelligence. This area emerges as a vision of the future information society which will combine computing, ubiquitous communication and user-friendly interfaces with support for interaction with humans.The aspects that may describe it also portray what one can expect from this area, i.e. to be sensitive, adaptive, intelligent, transparent, ubiquitous and responsive [3, 4]. Healthcare is arguably the field in which the development of ubiquitous systems is needed the most, since one has to make decisions based on immediate factual information [5]. Therefore, the health care sector represents a myriad of opportunities for the creation and implementation of this kind of systems. Given the current needs in terms of health care, it is essential to develop new support systems that are able to create safe and flexible environments in order to satisfy the needs of each patient. It is felt that in the future, all healthcare institutions will be equipped with systems capable of interacting with the environment [2].

2 An Ambient Intelligence Based Architecture for Patient Monitoring

This is a system that aims to ensure a continuous and effective patient monitoring, allowing precise location of labour and resources [6]. It is a multi-agent system that includes reasoning mechanisms and context-aware technologies. Thanks to its modular design, it can be integrated with other systems existing in the hospital. The proposed architecture is built on 3 (three) different modules, as it is depicted in Figure 1:

- User & Environment module: This module refers to the information obtained of patients, health professionals,resources and the entire environment that surrounds them.
- Monitoring Module: The monitoring module intends to guarantee an effective monitoring through the use of diverse sensors and context-aware technologies.

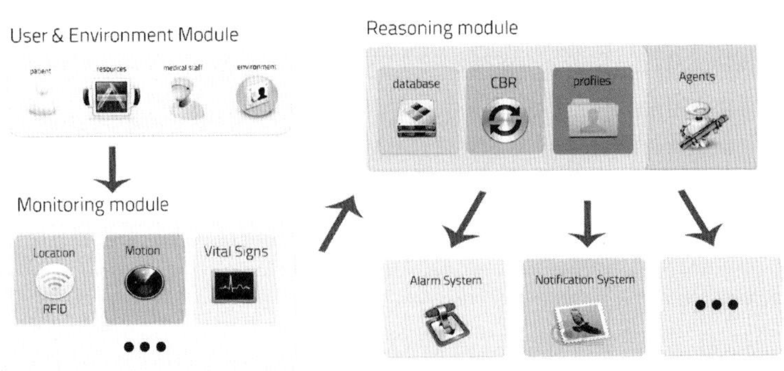

Fig. 1 The architecture for the monitoring system being proposed

- Reasoning Module: The main function of this module is to perform actions according to the conditions verified and it is responsible for managing the information gathered through the previous modules.

All patients, healthcare professionals and the resources are monitored and fully traceable through the use of RFID tags. When a patient moves from one room to one another, it triggers a notification to the doctor(s) that is(are) responsible for his/her condition. The same principle is applied to the resources when they are being used or have been transferred to other units.

3 The Architecture

The system proposed must be able to accept and integrate a very heterogeneous group of devices and technologies. This means that the architecture must provide communication and information mechanism where all different components can work together. To ensure this, it will be used two different service-oriented standards: OSGi and WebServices. Since in AmI there are multiple entities communicating with one another, providing and requesting services, it is easy to understand the interest in using such technology [1]. This means that the system proposed can be used by any other higher level architecture and use services provided by that architecture. It also results in a very expansible architecture as it is easy to add new components, which does not affect the components already present.

4 An Implementation in a Real Environment

Thefts of newborns are not a rare phenomenon and the number of new casualties rises every year [7]. Aware of this scourge, the administration of the Centro Hospitalar do Tâmega e Sousa, E.P.E. (CHTS) enforced procedures to guarantee the safety of newborns, one measure being the implementation of an RFID based location system. An RFID bracelet is attached to every newborn allowing their identification and location. Doctors, nurses or visitors also have an RFID based identification. All the events are registered on a database, which allows for a complete record of all activities being carried on.

Using logic based reasoning mechanisms for the data and a model of the different user profiles it would be possible to have several levels of alerts depending on whether it is a doctor, nurse or any other person. As more situations arise, the e-learning module has more cases to consider, therefore enhancing the ability of the system to react to unexpected situations. In this healthcare facility all the services report to the same entity, a computational framework called AIDA (an Agency for the Integration, Diffusion and Archive of information) which enables the interface between the existing systems and applications by using SOA (Subjective, Objective Assessment planning), establishing the communication through normalized protocols (like HL7). As represented in Figure 2, the reasoning module communicates

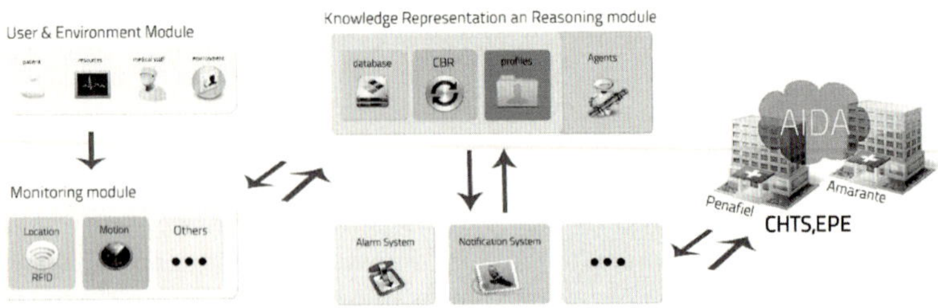

Fig. 2 Integration between the proposed system and the AIDA framework that runs in the CHTS-E.P.E.

with AIDA, and then may interact with other systems or simply uses the information directly received.

5 Conclusions

In this paper it were presented the building block of a framework for patient monitoring. It includes context-aware technologies that associated to knowledge representation and reasoning mechanisms (e.g. CBR) aims to create a safe, adaptable environment where patients, doctors and resources may be monitored at any time. With the utilization of open-standards of communication like OSGi and WebServices, the system may be expansible and easily integrated with other ones, which represents a gain in interoperability. We just started the implementation of the architecture proposed on the nursery unit of the CHTS - E.P.E., Portugal.

References

1. Cook, D.J., Augusto, J.C., Jakkula, V.R.: Ambient intelligence: Technologies, applications, and opportunities. Pervasive and Mobile Computing 5 (2009)
2. Sneha, S., Varshney, U.: Enabling ubiquitous patient monitoring: Model, decision protocols, opportunities and challenges. Decision Support Systems 46(3), 606–619 (2009)
3. Moreno, A., Nealon, J.L.: Applications of Software Agent Technology in the Health Care Domain. Birkhauser (Architectural), Basel (2004)
4. Lanzola, G., Gatti, L., Falasconi, S., Stefanelli, M.: A framework for building cooperative software agents in medical applications. Artificial Intelligence in Medicine 16(3), 223–249 (1999)
5. Costa, Â., Novais, P., Costa, R., Machado, J., Neves, J.: A memory assistant for the elderly. Intelligent Distributed Computing III, 209–214 (2009)
6. Carneiro, D., Novais, P., Costa, R., Gomes, P., Neves, J.: Emon: Embodied monitorization. In: AmI 2009: Proceedings of the European Conference on Ambient Intelligence, pp. 133–142. Springer, Heidelberg (2009)
7. N.C. for Missing and E. Children, 2008 annual report (2009), http://www.missingkids.com/en_us/publications/nc171.pdf

Helping Elderly People and Persons with Disability to Access the Information Society

Diego Gachet, Manuel de Buenaga, Víctor Padrón, and M. Alanbari

Abstract. NAVIGA is an European project whose main goal is to design and develop a technological platform allowing elderly people and persons with disability to access the Internet and the Information Society through an innovative and adaptable navigator. NAVIGA also allows the creation of services targeted to social networks, mind training and personalized health care.

Keywords: elderly, wellbeing, ambient assisted living.

1 Introduction

Nowadays elderly and disabled people have serious difficulties to access the information society and the benefits it provides in areas such as eHealth or leisure. Older people often aren't comfortable driving a computer and the technological devices seems to be complicated for them. This problem may worsen due decreasing of cognitive, visual, hearing or motor capacity. The Naviga project (2009-2012) is an European effort funded under Eurostar [1] program and whose main objective is to provide these collective simple methods to overcome these barriers and enjoy greater personal autonomy and a better quality of life. To do this, the Naviga project will provide an integrated platform that facilitates an Internet access for all through their computer or TV, and also allowing employ alternative keyboard for navigation devices.

In addition, the Naviga platform will have services that facilitate the incorporation of elders and people with diverse functional capacity to the Information Society using Information and Communication Technology to maintain their minds agile and

Diego Gachet · Manuel de Buenaga · Víctor Padrón · M. Alanbari
Escuela Politécnica Superior
Universidad Europea de Madrid
C/ Tajo s/n, 28670 Villaviciosa de Odón, Madrid, Spain
e-mail: gachet@uem.es, buenaga@uem.es,
victor.padron@uem.es, alanbari@uem.es

J.C. Augusto et al. (Eds.): ISAmI 2010, AISC 72, pp. 189–192.
springerlink.com

involve them actively in his community of relatives and friends, while allowing them to control their health. Moreover, as an open system, the Naviga platform allows connecting external devices, so users can connect various accessible devices as controls, buttons, or specific sensors. Similarly the Naviga platform allows the incorporation of new services as well integrating the existing ones and provides a common interface for interaction with users.

2 Research and Development in Naviga

The research and development part of the NAVIGA project will be based on the *Reference Model of Open Distributed Processing* (RM-ODP)[2]. The RM-ODP framework was a joint effort by the international standards bodies ISO and ITUT to develop a coordinating framework for the standardization of open distributed processing. The model describes an architecture within which support of distribution, interworking, interoperability and portability can be integrated. With the aid of RM-ODP it will be easier to organize the pieces that compose NAVIGA into a coherent whole. RM-ODP defines five viewpoints (providing concepts, structures and rules) from which to abstract ODP systems. The main advantage of RM-ODP is that specifying NAVIGA using these viewpoints allows an otherwise large and complex project to be separated into manageable pieces, each focused on the issues relevant to different members of the development team. The viewpoints defined by RM-ODP are: Enterprise: purpose, scope and policies. Information: information and information processing. Computational: functional decomposition. Engineering: infrastructure required to support distribution. Technology: choices of technology for implementation. Based on the RMODP viewpoints, the work to be performed in NAVIGA project has been organized into several phases and tasks within the phases.

3 Naviga Main Components

Multi-functions platform
The system will be used for specialized teleconsulting, for memory and nutritional reports, for tele-reeducation at home, for diagnostic tele-assistance or therapy, for tele-surveillance with detection and analysis of abnormal reactions.

Adaptable device and communications interfaces
One of the task of the Naviga project focuses on the development of a complete intelligent platform allowing data acquisition / reception / processing / communication: audio scene analysis and robust voice recognition, image and pattern recognition, people falling and alarm system could be integrated to this intelligent platform.

Web navigator
In the case of web navigator the way most people 65+ get internet information is via web browser, the problem therefore is to improve the interaction way between elderly people and browser via inputs elements as pulsers or voice commanded

interfaces, and at the other hand adapting web pages providing information that senior needs [3]. In this sense NAVIGA introduce innovation at the design level of web browser applying the "Design for All" concept.

Social networks
NAVIGA will permit to create social networks. We have recently seen on Internet the appearance of the Web 2.0, the second generation of web based communities and services (wikis, folksonomies, blogs, mashups)[4] with the aim of facilitating the collaboration and sharing of contents between users.

Healthcare services
NAVIGA will ease the development of Healthcare services in a multi-functional integrated platform.

Stimulation and entertainment systems
The value proposition includes enhanced memory, delayed cognitive decline (and physical), improved quality of life, reduced caregiver burdens and reduced cost of care to payers. These include computer-based cognitive stimulation products that rely on the plasticity property of memory[5]. There is preliminary evidence that these technologies may have positive impacts on memory in the short term. NAVIGA innovates incorporating embedded assessment capabilities.

4 Expected Results

The expected results of the project from the development point of view are: a hardware interface device set adaptable to the senior people and persons with disabilities and enabling the interface with the computer or television. A framework (tools and methods) for the creation and deployment of services and applications. The development of services including a web navigator allowing the access of elderly people and persons with disabilities to the Internet. Two technology demonstrators in the area of eHealth and Entertainment. An analysis of business opportunities and business requirements (identifying their key strengths and weakness) for the successful commercialization of project results.

Naviga project includes two case-study / scenarios to demonstrate the functionality of the developed framework. A concrete pilot application will be developed and evaluated in a Hospital targeting at the validation of a medical care service. Another scenario will be developed and evaluated in a specific Healthcare centre, targeting the access of elderly people and persons with disabilities to the Information Society through web specific navigator access, social networks and mind training. The scenarios will include the participation of real end-users to validate the technological developments. The consortium is composed of 5 research-performing SMEs, Answare (ES), Encore (ES), Entropy (ES), Mind Autonomy Research MAR (FR) and Technosens (FR), investing min. 20% of annual turnover in R&D, 1 University and two end users (an Hospital and a daily Healthcare centre located both in the Madrid region) will participate in the project.

5 Conclusions

The results of the Naviga Project will be ensuring from its early stages with the support of various end users allowing a correct project requirements stage, based on the real needs of the groups involved, and participating in the process of validation interfaces and services. End users with diverse functional capacity will test and validate project services us to know if them are useful. The Naviga project brings together six European entities, all experts in various areas of social and ICT sectors.

Acknowledgments. Naviga is an Eurostar project partially funded in Spain by Centro para el Desarrollo Técnico Industrial (CDTI) under the support program for sme participation in international programs.

References

1. Eurostar program web site (2010), http://www.eurostars-eureka.eu/ (accessed June 24, 2009)
2. RM-ODP website (2010), http://www.rm-odp.net/ (accessed June 2, 2009)
3. Clerk Maxwell, J.: A Treatise on Electricity and Magnetism, 3rd edn., vol. 2, pp. 68–73. Clarendon, Oxford (1892)
4. Gartner, Gartner Reveals Five Social Software Predictions for 2010 and Beyond (2010) http://www.gartner.com/it/page.jsp?id=1293114 (accssed December 12, 2009)
5. Czaja, S.J., Lee, C.C.: Information technology and older adults. In: Jacko, J.A., Sears, A. (eds.) The human-computer interaction handbook, 2nd edn., Erlbaum, New York (2007)
6. Boyd, D.M., Ellison, N.B.: Social network sites: definition, history, and scholarship. Journal of Computer-Mediated Communication 13, 11 (2007)

Context Triggered Experience Sampling

David V. Keyson

Abstract. As products and services have become more embedded in the everyday life and routines of users, the need to understand how the context of use influences product usage over time has increased. The Experience Sampling Method (ESM) aims to capture both the context and content of the daily life of individuals. Critical to the ESM methods is the notion of asking the appropriate set of experience related questions at the right moment. The current approach to ESM is based on provided the user with a set of qualitative questions, typically using scaled measures, at pre-defined intervals during the course of the day over a period of several days or longer. In the current paper the Context Aware Toolbox (CAT) is described as a means to rapidly equip an environment with sensors and actuators which in turn can trigger user input via formal questionnaires and indirect input based on the task at hand. The main elements of the CAT is a database of algorithms, written in MAX MSP which can be rapidly adapted to particular needs, Phidget sensors and actuators, a MAX Mini containing instructions and the database, touch screens for displaying the qualitative measures on site, and installation tools, as well as a web based infrastructure for adding new content to the database. The CAT setup is currently being applied to design research cases in the area of social and environmental sustainability; an application design case focused on sustainable living is described.

Keywords: Experience Sampling, Context Awareness, Prototyping.

1 Introduction

The emergence of products and services embedded in the everyday routine of people reflect the degree to which technology has become a part of lives. Increasingly, the only way to evaluate the usability of products and services, behind the immediate ease of use of the interface, is to study user interaction in context. For pure

David V. Keyson
ID-StudioLab, Fac. of Industrial Design Engineering, Delft University of Technology, Landbergstraat 15, 2628 CE, Delft, The Netherlands
e-mail: d.v.keyson@tudelft.nl

J.C. Augusto et al. (Eds.): ISAmI 2010, AISC 72, pp. 193–196.
springerlink.com

Internet based applications, this might mean launching a "beta" version of the software and observing what happens with large number of users and then fixing the problems over successive beta releases. In the case of evaluating physical products in context, focus group sessions, or phone or mail based questionnaires may only provide brief glimpses into product adoption and usability issues. The two predominant methods to gather ongoing usage data are the Diary Recollection Method [1] and the Experience Sampling Method [2]. Both methods can also be combined [3]. In the case of DRMS the user may be given a diary to fill-in over a fixed time interval across several days or more, or maybe given a camera to take pictures of the context of use on a regular basis. Experience sampling may be conducted via asking the user to fill in a paper form of electronic form on the basis of timed requests which may generated via a watch, beeper, Website, or PDA, to name a few possible platforms. It is generally argued that ESM combines the ecological validity of naturalistic behavioral observations with the nonintrusive nature of diaries and the precision of scaled questionnaire measures. While the limitations of ESM are acknowledged, in particular the demands it imposes on respondents, which may contribute to a self-selection bias and selective non-response, the method is regarded as some as the best alternative to capturing the context of use. Central to the ESM is the notion that "by sampling experience the moment it occurs, it avoids the potential distortions associated with the use of daily or weekly retrospective diaries" [2]. While this seems at first glance a logical statement, taking a deeper look at ESM studies reveals that often ESM based questionnaires appear at pre-set time intervals, but are not necessarily triggered when the behavior under study occurs in a given context. Furthermore the ESM assumes that experiences can only be recorded via qualitative scales, in some cases the way in which a product or service is engaged may in itself reveal experiential data. For example the degree to which a user applies force to while setting a physical control may be an indication of stress or the number of repeated efforts to make a "soft" selection on a screen could reflect a degree of confusion. Such data could be monitored in real-time and collected over an extended period. Taking things a step further the product or service itself could be remotely updated given the availability of online usability data. In some cases it may make sense deploy sensors and actuators close to the product or additionally in the vicinity of the product to capture experience data. To this extent the Context Aware Toolkit was developed.

2 Context Aware Toolbox

The Context Aware Toolbox (CAT) provides a means to rapidly equip environment with sensors and actuators to enable the monitoring of activates such that they can trigger an experience sampling survey or that that the actuators and sensors themselves can be used to collect data. For example could sensors could be placed on the taps of a sink, such that when the home dweller first turns on the hot water and than adds cold to reach the right temperature, which generally results in a waste of hot water compared to the reverse sequence, a short questionnaire to determine why this behavior was followed could be displayed. The results of this experience sampling could then be used to generate design insights leading to a redesigned faucet that encourages the right order. Sensor technology can also be used to trigger video recording of behavior when using the faucet. For example a video system, with a narrow view camera, could be coupled to the sensors to record the specific hand actions at moment of faucet use.

Table 1 Main Components of the Context Aware Toolkit

Sensors	IR distance, light, force, rotation, touch, humidity, motion, pressure, temperature, orientation, vibration, webcam
Switches	micro switches, switch pads
Actuators	LEDs, servomotor, speakers, text LCD
Phidgets	Phidget interface kit 8/8/8, PhidgetRFID kit, Phidget Accelerometer 3-axis, Phidget Encoder.
Tools	Soldering iron, pliers, cables, wires, scissors
Computing	Mac Mini with Max/MSP and component database, experience sampling example scales, compact touch screens with clip-on fast mounting system

The main components of the CAT are sensors, switches, actuators, Phidgets, Tools and Computing modules as depicted in Table 1 below. The CAT supports the flexible setup of activity monitoring and experience sampling by a visual step-by-step operation guide (Figure 1).

Fig. 1 The Context Aware Toolbox

The toolbox database provides easy access to many sample project implementations (i.e., patches) developed in Max MSP and provides the researcher with a table of available sensors and actuators as well as ordering information, should they not be present in the toolbox. The database was built in FileMaker Pro and runs on a Mac Mini that is the lower drawer of the toolbox. The database is web compatible supporting remote access and development.

In order to organize and access the stored material in a convenient way a website structure was designed. The website gives direct functional support (e.g., tutorials, database and visuals) The website can also support the addition of new technology components, example projects and designer experiences. Sample layouts and scales for experience sampling are also provided in the database.

3 Discussion and Future Work

The Context Aware Toolbox is currently being developed to support studies on sustainable living. Modern sustainable homes are often rather high-tech environments. For example, heating, cooling, ventilation and energy management systems can all interact with each other. Home occupants should be able to monitor, manage, maintain such systems. By enabling experience sampling to occur at the right place and time, the expectation is that deeper insights into product use and routines over time will be gained. The importance of understanding how products fit into routines has been exemplified in the work by Shove [4] on practice based theory, which suggest that the "things" people use should be considered in terms of how they fit into grounded routines. Insights into the context of use can help support the development of smart products that are context aware.

In conducting in-context studies across multiple homes, various configurations of sensors and actuators will be required to adapt to the particular context. To this extend the modular configurability and portability of the CAT is expected to be of great benefit.

Acknowledgements. The work presented in this paper was part of the Design for Wellbeing project. The author gratefully acknowledges the support of the Innovation-Oriented Research Programme 'Integral Product Creation and Realization (IOP IPCR)', funded by the Dutch Ministry of Economic Affairs.

References

1. Kahneman, D.m., Krueger, A.B., Schkade, D.A., Schwartz, N., Stone, A.A.: A Survey Method For Characterizing Daily Life Experience: The Day Reconstruction Method. Science 306, 1776 (2007)
2. Hektner, J.H., Schmidt, J.A., Csiksaentmilayi, M.: Experience Sampling. Sage Publications, Thousand Oaks (2007)
3. Khan, V.J.: Mediated Awareness For Intral-Family Communication. PhD Dissertation, Eindhoven University of Technology (2009)
4. Shove, E.: The Design of Everyday Life, Berg (2007)

Adaptive Experience Sampling: Addressing the Dynamic Nature of In-Situ User Studies

Martijn H. Vastenburg and Natalia Romero Herrera

Abstract. Experience sampling techniques can be used to study user experiences with products in a natural setting and over time. Nowadays, researchers can use selective sampling to link the timing and questions to relevant product events and contextual events. Existing research has focused on maximizing the quality and quantity of feedback, while at the same time minimizing interruptions and maintaining the motivation of the participants. In practice, however, the optimal timing and content of questions also depends on the changing interests of the researcher, a topic that has not yet been addressed. This position paper introduces 'adaptive experience sampling', i.e. a method that enables researchers and designers to change the focus of their experience sampling study on the fly. Using adaptive experience sampling, researchers and designers can better link the sampling to their gradually growing insights.

Keywords: Experience Sampling, In-Situ Research, Evaluative Tools.

1 Introduction

Evaluation of ubicomp products tends to be challenging. Design concepts tend to be radically different from traditional interactive products, and interaction tends to be embedded in everyday life. Feedback from users is needed to evaluate design directions and to iteratively improve design concepts. Traditional evaluation methods, which are valuable when assessing product use in an isolated setting, are inadequate for studying product usage and user experience in relation to everyday activities. User experiences can be highly dynamic and evolve along with the use

Martijn H. Vastenburg . Natalia Romero Herrera
ID-StudioLab, Fac. of Industrial Design Engineering, Delft University of Technology, Landbergstraat 15, 2628 CE, Delft, The Netherlands
e-mail: {M.H.Vastenburg,N.A.Romero}@tudelft.nl

J.C. Augusto et al. (Eds.): ISAmI 2010, AISC 72, pp. 197–200.
springerlink.com

of a product or services. An evaluation methodology needs to capture the dynamics of the environment as well as the changes in user experience caused by user-product interaction. At the same time, the user experience could also be affected by external factors; for example the user experience of a social network product could be affected by social interactions external to the product. Longitudinal field studies and subjective analysis are therefore needed to capture how people experience the product concepts, and to understand how a product fits in everyday life.

The experience sampling method (ESM), a research methodology developed by Larson and Csikszentmihalyi [3], was designed to capture user experiences in the field. Initially ESM took advantage of the popularity of earlier mobile devices (e.g. pagers) to ask people for feedback at random times during the day. This configuration aimed to reduce problems that participants might have when recalling events, a problem underlying many self-report techniques. With ESM participants make a quick record close to the moment of interest, rather than having to recall what they did in the past.

As mobile technology evolves, the sampling process becomes more intelligent and sensitive to the context of the product use. Nowadays, researchers can use selective sampling [5, 2], a sampling technique that links the timing and questions to relevant events. Figure 1 shows a typical configuration of selective ESM, in which a portable device is used to collect user feedback. The ESM controller uses sensor data that could capture contextual as well as user-product events to select relevant sampling moments. In addition, the controller, based on the same information, may decide what question or flow of questions should be asked and how should be presented together with the format of the answers.

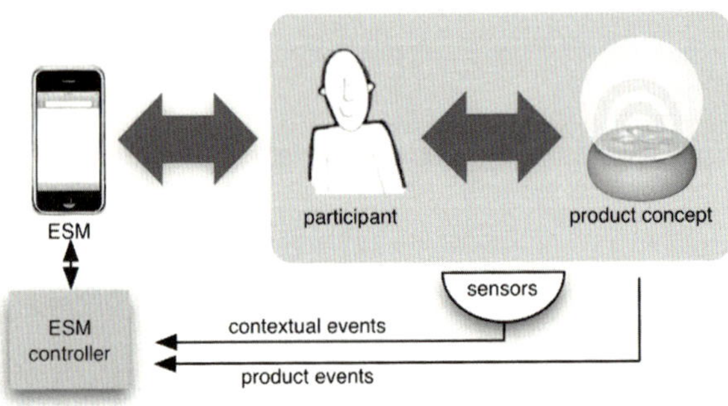

Fig. 1 Typical configuration of selective sampling. The user experience of a participant interacting with a product concept is being studied. Product events and contextual events are used to activate questions on a portable device.

A common challenge when using ESM is to maximize the quality and quantity of the samples while minimizing interruptions and maintaining the motivation of the participants. The development of strategies to optimize sampling, interruptions and motivation has been the key focus in recent ESM research [e.g., 1, 4, 6, 7, 8].

Using these strategies, timing and content of questions can be adapted to the actual state and context of the participants and the product.

The possibility to digitally collect data in-situ opens opportunities for researchers to instantly gain insights in users and product behavior, which might lead to new or improved research questions, thereby changing the initial focus of the study. In recent field studies using ESM, we have experienced problems adapting ESM to preliminary findings from the field study. For example, when a researcher finds out that certain questions are not being asked, it can be difficult to change the configuration. Researchers need to be able to easily inspect results and to adapt the timing and content of questions in the course of a study. Changing research interests in ESM have not been addressed before. The present position paper introduces the adaptive experience sampling method to include mechanisms that support researchers in adapting timing and content of experience sampling on the fly.

2 Adaptive Experience Sampling

The key assumption underlying adaptive experience sampling is that researchers cannot know all questions nor the detailed product use and context of use in advance. Adaptive ESM (A-ESM) aims to enhance existing experience sampling methods by providing instruments that enable researchers to easily inspect the preliminary findings in relation to context and product usage data, and to easily adapt the questions and the triggers (Figure 2).

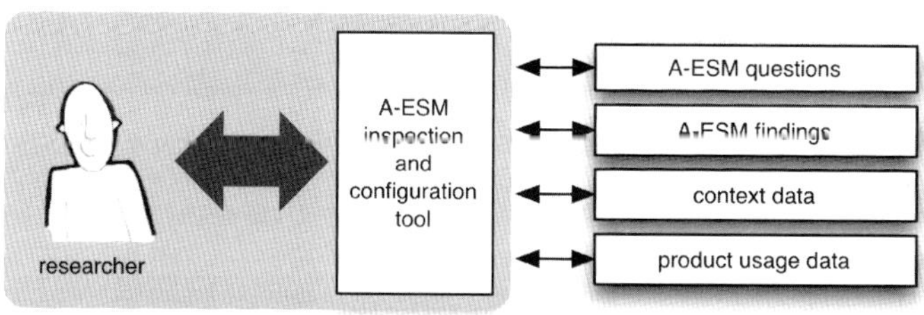

Fig. 2 Adaptive experience sampling configuration. Since researchers cannot know all questions nor the detailed product use and context of use in advance, tools are needed which enable researchers to easily inspect preliminary findings and adapt the questions.

Tools to support adaptive experience sampling are now being developed. The key design challenges are: (1) researchers should be able to detect unexpected behaviors, (2) researchers should be able to detect that anticipated behaviors do not occur, and (3) researchers should be able to view the findings of their experience sampling study in relation to the configuration in time. At the same time, the tools should enable researchers to easily change the behavior of sensors (e.g., changing thresholds and sampling frequency), define triggers and adapt questions.

3 Discussion and Future Work

In this paper we want to position our work on adaptive experience sampling, stating that current ESM studies could be improved by enabling researchers to adapt the ESM protocol according to changes in context of use, user behavior and research interest. The explorative stages of the product design process require a dynamic experience sampling approach. Using adaptive ESM, researchers and designers can use experience sampling methods in a way which better addresses their changing interests as compared to traditional ESM methods. The proposed A-ESM should therefore be considered as a valuable addition to the existing ESM methods. Tools and instruments which support researchers and designers in using the A-ESM methodology are now being developed. Case studies are planned to evaluate and improve the approach in different settings to cover a wide range of user, product and researcher needs.

Acknowledgements. The work presented in this paper was part of the Design for Wellbeing project. The authors gratefully acknowledge the support of the Innovation-Oriented Research Programme 'Integral Product Creation and Realization (IOP IPCR)' of the Dutch Ministry of Economic Affairs.

References

1. Consolvo, S., Walker, M.: Using the experience sampling method to evaluate ubicomp applications. IEEE Pervasive Computing 2(2), 24–31 (2003)
2. Consolvo, S., Harrison, B., Smith, I., Chen, M., Everitt, K., Froehlich, J., Landay, J.: Conducting in situ evaluations for and with ubiquitous computing technologies. Intl. Journal of HCI 22(1), 107–122 (2007)
3. Hektner, J.M., Schmidt, J.A., Csikszentmihalyi, M.: Experience sampling method: Measuring the quality of everyday life. Sage, CA (2007)
4. Hsieh, G., Li, I., Dey, A., Forlizzi, J., Hudson, S.E.: Using visualizations to increase compliance in experience sampling. In: UbiComp 2008, pp. 164–167. ACM, New York (2008)
5. Intille, S.S., Tapia, E.M., Rondoni, J., Beaudin, J., Kukla, C., Agarwal, S., Bao, L., Larson, K.: Tools for studying behavior and technology in natural settings. In: UbiComp 2003, ACM, New York, pp. 157-174 (2003)
6. Intille, S.S., Kukla, C., Ma, X.: Eliciting user preferences using image-based experience sampling and reflection, pp. 738–739. ACM, New York (2001)
7. Kapoor, A., Horvitz, E.: Experience sampling for building predictive user models: a comparative study. In: CHI 2008, pp. 657–666. ACM, New York (2008)
8. Khan, V., Markopoulos, P., Eggen, B., IJsselsteijn, W., de Ruyter, B.: Reconexp: a way to reduce the data loss of the experiencing sampling method. In: MobileHCI 2008, pp. 471–476. ACM, New York (2008)

An Intelligent Domotics System to Automate User Actions

Natalie Kcomt Ché, Niels Pardons, Yves Vanrompay,
Davy Preuveneers, and Yolande Berbers

Abstract. Home automation systems are ambient intelligence systems that are designed to help people proactively, but sensibly. In this paper we propose a system that learns and automates patterns in the interactions of the user with the home automation devices. We will show our approach and architecture. We use an event processing tool to handle the events from the home automation devices, prediction algorithms to predict the next action and reinforcement learning to decide which action are suitable to be automated.

1 Introduction

Today many homes are filled with home automation devices and sensors. In the future these devices will work together to form ambient intelligence (AmI) systems. A possible definition for an AmI system is as follows: a digital environment that proactively, but sensibly, supports people in their daily lives [2].

A person who lives in a house often uses the same devices every day and in the same order. The goal is to learn, recognize and automate these patterns of interactions with the various home automation devices. A classic example of such a pattern is to wake up, turn off the alarm clock and then make coffee. An intelligent learning system could learn this pattern and automatically make coffee in the morning, when the person turns off the alarm clock. The main assumption by our system is that

Natalie Kcomt Ché · Niels Pardons · Yves Vanrompay · Davy Preuveneers · Yolande Berbers
Katholieke Universiteit Leuven, Dept of Computer Science Celestijnenlaan 200A 3001 Heverlee, Belgium
e-mail: `natalie.kcomtche@student.kuleuven.be`,
`nielsroger.pardons@student.kuleuven.be`,
`Yves.Vanrompay@cs.kuleuven.be`,
`Davy.Preuveneers@cs.kuleuven.be`,
`Yolande.Berbers@cs.kuleuven.be`

J.C. Augusto et al. (Eds.): ISAmI 2010, AISC 72, pp. 201–204.
springerlink.com

humans follow such patterns. This has been demonstrated already in related work. When there is small noise, because of the users not following patterns, the system should not automate the action and should also not learn this noise.

The main challenge for such a system is to predict and automate enough actions in a timely fashion, avoiding that the user manually has to start these actions, while preventing to automate actions that the user does not want.

The following is a possible user scenario for our system. "Somewhere in a house in the future Sofie wakes up and turns off the alarm clock. The curtains open automatically. Sofie gets up and goes to the bathroom. The light turns on automatically as soon as she enters. The system has learned that she always showers in the morning and the water is heated to the preferred temperature. After the shower Sofie goes to the kitchen where her black coffee is already made. The system has learned that she prefers black coffee in the morning."

In section 2 we explain the architecture of our system. Related work is discussed in section 3. Conclusions and future work follow in section 4.

2 The Intelligent Domotics System

We adopt the principles of SOA for the creation of the services provided by the various domotic devices and sensors in the environment. Furthermore, the principles of Event Driven Architecture (EDA) are incorporated as they are used for systems that send events between independent software components and services. Events are generated by domotic devices and sensors in the environment and the learning system must respond. Our architecture uses a combination of EDA and SOA, called event driven service oriented architecture (ED-SOA)[3]. Here EDA is used to expand SOA with the publish-subscribe communication pattern and the ability to collect events over a long period of time, after which they can be processed and correlated. SOA is needed because the domotic devices provide services to each other and to the user. EDA handles the events that the devices generate asynchronously and that need to be collected over a long period of time.

Figure 1 shows the design of the learning domotics system, consisting of four different layers. The physical layer contains the various sensors and home automation devices. The communication layer provides communication between the different devices. The information layer retrieves information and higher level abstraction from the sensors and home automation devices. Finally, the decision layer controls the devices. We will first discuss the information layer and then the decision layer.

The information layer includes the event processing tool, Esper, which receives, processes and forwards the events to the other components. Esper can do filtering, but could also aggregate events to represent more hierarchical or composite events. The prediction algorithms in this layer will predict the next action based on recent history. We will use the Jacobs-Blockeel algorithm[6] and the FxL algorithm[5] for the prediction. Jacobs-Blockeel is based on IPAM, which uses a first order Markov model. Jacobs-Blockeel uses a mixed order instead to calculate the probability distribution for next event. Initially first order Markov models are used. Whenever the

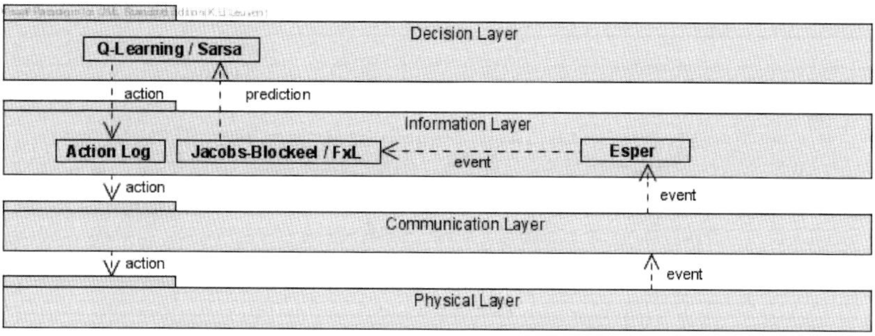

Fig. 1 The design of the learning system

algorithm makes a correct prediction, a higher order is activated, otherwise the order does not change. For example if the system correctly infers the rule that after observing 'a', 'b' will be observed then the system treats 'b after a' as a single observation. The advantage of using a mixed order Markov model is that high orders are only used when necessary, which benefits the storage and processing requirements. Jacobs and Blockeel claim that the highest order for the Markov model is not always the best choice to determine the probability for the next event. We chose for the Jacobs-Blockeel algorithm because in a home automation environment there are lots of possible actions to take. Using a system such as IPAM that only takes the most recent action into account is not sufficient here. Furthermore the Jacobs-Blockeel algorithm but mixes orders intelligently treating frequently occurring sequences as single observations

FxL is an on-demand approach for combining the results of different order Markov models. It is based on an n-gram tree that contains the frequencies of different input subsequences with a length up to a specified value k. These n-gram models assign a score to each symbol, which represents the probability that a symbol appears next in the input sequence. This relatively new algorithm yields very good results in the context of predicting commands in a Unix terminal. FxL was chosen because it is able to get these good results while keeping the storage costs limited by the specified k and the amount of possible user actions, whereas Active LeZi, with similar results, has storage costs that grow with the dataset size.

The decision layer contains the decision algorithms Q-learning and SARSA to determine when to execute which action. The prediction of the algorithm is recorded in the state of the Q-learning algorithm. The Q-learning algorithm is a reinforcement learning algorithm that is able to compare the different actions to take without explicitly modeling the whole environment with very good results. SARSA is a variant of Q-learning that does not make the assumption that the optimal policy can be followed. The final action to be executed is sent by the Q-learning algorithm to the home automation device in question. This layer works on top of the prediction system as an extra precaution to protect against possible errors the prediction system might make.

3 Related Work

MavHome[1] uses the Active Lezi (ALZ) prediction algorithm and its accuracy is improved by using frequent patterns discovered through Episode Discovery (ED). The decision algorithm is TD(0) reinforcement learning. The iDorm[4] system extracts fuzzy membership functions and fuzzy rules from the user data to model the behavior of the user and employs a fuzzy logic controller to determine which action to automate. The Adaptive House[7] uses artificial neural networks to predict the next state of the house. It balances energy cost and discomfort cost using Q-learning to determine which action to automate. The design of our system is inspired by MavHome. Our system differs from MavHome in the use of an event processing tool to process events and the use of other learning techniques: ED is not employed and Jacobs-Blockeel and FxL are used instead of the ALZ algorithm. Furthermore, a variant of Q-learning, SARSA, is implemented as the decision algorithm.

4 Conclusion

We proposed an intelligent domotics system that learns patterns in the interactions of the user with the home automation devices and then automates these interactions. The architecture of the system is a combination of EDA and SOA, also called ED-SOA. For learning and automating user actions, we apply various machine learning techniques: Jacobs-Blockeel, FxL, Q-learning and SARSA. These algorithms decide what action should be performed. Concerning future work, the system will be implemented with these different algorithms, and be tested and evaluated with both synthetic and real data from MavHome. We will compare the performance of the two prediction algorithms and of the two decision algorithms. Finally, we will compare our system with other existing projects.

References

1. Cook, D.J.: Mavhome: An agent-based smart home. In: PERCOM 2003, p. 521 (2003)
2. Cook, D.J.: Review: Ambient intelligence: Technologies, applications, and opportunities. Pervasive Mob. Comput. 5(4), 277–298 (2009)
3. Ghalsasi, S.Y.: Critical success factors for event driven service oriented architecture. In: ICIS 2009, pp. 1441–1446 (2009)
4. Hagras, H.: Creating an ambient-intelligence environment using embedded agents. IEEE Intelligent Systems 19(6), 12–20 (2004)
5. Hartmann, M., Schreiber, D.: Prediction algorithms for user actions. In: LWA, pp. 349–354 (2007)
6. Jacobs, N., Blockeel, H.: Sequence prediction with mixed order markov chains. In: Proceedings of the Belgian/Dutch Conference on Artificial Intelligence (2003)
7. Mozer, M.: The neural network house: an environment that adapts to its inhabitants. In: Proceedings of the AAAI Spring Symposium on Intelligent Environments, pp. 110–114 (1998)

Assessing the Usefulness and Acceptance of HERMES MyFuture System in Two European Countries

Cristina Buiza, Ana Belén Navarro, Mari Feli Gonzalez,
Arjan Geven, Manfred Tscheligi, and Sebastian Prost

Abstract. In this paper, the results of the first user trials of the HERMES system "Cognitive Care and Guidance for Active Aging", a research project co-funded by the European Union under the Seventh Framework Program (FP7) are presented. In particular it describes how elderly users in Austria and Spain tested HERMES MyFuture, which aims to address prospective memory problems of elderly people by means of setting different kinds of appointments (time based, location based). In general, the system is perceived as useful and easy to use.

Keywords: Usefulness, Assistive Technology, Elderly, Aging.

1 Introduction

Usefulness and user acceptance of technology plays a critical role in the success of the Assistive Technology [1]. User acceptance can be defined as the willingness of a certain user group to employ a certain system for the task it was designed to support [2]. Social and cultural background, attitudes and values strongly influence how a technical tool is perceived. For this reason, it is essential to know

Cristina Buiza · Ana Belén Navarro · Mari Feli Gonzalez
Fundación Instituto Gerontológico Matia – INGEMA. Camino de los pinos 27, 20018. Donostia – San Sebastián. Spain
e-mail: {cristina.buiza,ana.navarro,mari.gonzalez}@ingema.es

Arjan Geven · Manfred Tscheligi · Sebastian Prost
Center for Usability Research and Engineering,
CURE, Hauffgasse 3-5, A-1110 Vienna, Austria
e-mail: {geven,tscheligi,prost}@cure.at

J.C. Augusto et al. (Eds.): ISAmI 2010, AISC 72, pp. 205–208.
springerlink.com

how elderly users interact with such a system in order to evaluate its performance and acceptance [3].

Older adults of this generation are reluctant to use external devices and aids that may reduce their autonomy or minimize their cognitive or functional efforts [4]. Technology plays a minor role in their life. New technologies can frighten elderly users, so they do not dare to even try things out. Elderly users can also simply ignore new technologies because they do not explicitly see how technology could help them or because they think they are not skillful enough. However, they are open to trying the use of new technologies. In this sense, external technological aids should be easier and simpler than the aids currently used.

"HERMES: Cognitive Care and Guidance for Active Aging" is a project co-funded by the European Commission within the 7th Framework Program (grant agreement number: 216709) [5]. HERMES has been developed with the aim of reducing age-related cognitive decline. It provides assistance but also promotes the autonomy and independence of users in their daily lives, employing pervasive, non-obtrusive technology. The main aims of the HERMES project are facilitation of episodic memory, reminding users of advanced activities and cognitive training.

In order to design adapted technology to mental and behavioural status of the users, it is necessary to test it with real users. In this paper, a part of the user evaluation of the first prototype is presented. Specifically, HERMES MyFuture (see Figure 1). This application, which is included in HERMES, supports users in planning their lives, so it aims to give support in prospective memory. MyFuture enables users to set appointments of different kinds (time-based, location-based), create to do lists and store notes for future use both at home and while on the go devices. Notes might be constituted of recorded audio or typed words. Also, MyFuture application has a mobile extension (HERMES PDA) which is used for recording appointments when the users are not at home. HERMES automatically synchronizes the PDA with the home-system. All future appointments will be available on both applications. Tasks with HERMES MyFuture cover browsing through scheduled events, introducing a new event and modifying an event. All these tasks were evaluated.

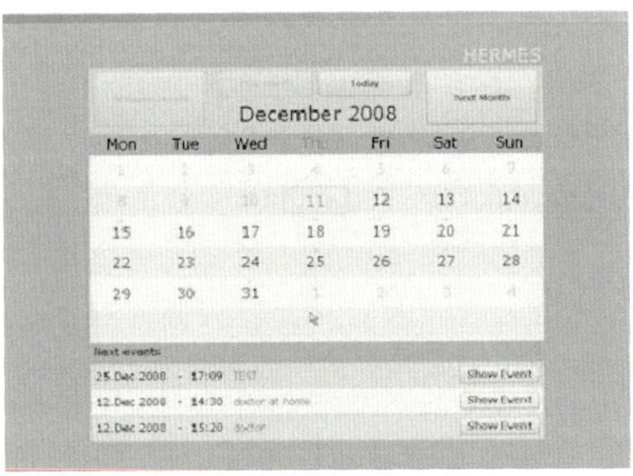

Fig. 1 Main screen of "HERMES MyFuture"

2 Methodology

The MyFuture application was evaluated in a lab environment in Spain and in Austria. In Austria 13 potential HERMES users participated (5 males and 8 females, with an average age of 71.92, from 62 to 86 years old) and in Spain, 14 (7 males and 7 females, with an average age of 69.57, from 63 to 76 years old).

The older users evaluated the HERMES MyFuture application with a mixed questionnaire (open and closed questions). The application was also evaluated using a questionnaire called *TAQ (Technology Acceptance Questionnaire)* based on the Unified Theory of Acceptance and Use of Technology (UTAUT). This questionnaire is a recent instrument, which consists of six constructs: Performance expectancy, Effort expectancy, Attitude towards using technology, Facilitating conditions, Anxiety and Behavioural intention to use the system. All items were measured on a 7-point Likert scale (from completely disagree to completely agree) [6].

3 Results

Regarding *HERMES MyFuture TAQ*, the majority of Austrians and Spanish participants had a good *Performance Expectancy*, that is, they believed that using HERMES MyFuture can be useful in their lives. They believed that it is rather easy to use (*Effort Expectancy*), they had the necessary resources to use it (*Facilitating Conditions*), and its use seems a good idea (*Attitude towards using technology*). Furthermore, they did not feel fear or anxiety at the time of use (*Anxiety*) and would not mind using it if they had access to it (*Behavioural intention to use the system).* Mann-Whitney Tests show significant differences in MyFuture between Spain and Austria in only one item: "A specific person (or group) is available for assistance with HERMES MyFuture difficulties" ($p=0.01$). Spanish older users are more in agreement with it than Austrian users ($M=17.75$ and 9.96, respectively). Also, there were gender differences in only one item: "*It scares me to think that I could lose a lot of information using HERMES MyFuture by hitting the wrong key*" ($p<0.01$). Older men are more in agreement with it that older women ($M=17.62$ and 11.10, respectively). However, there were no significant age differences (between users of up to 69 and 70+ years of age).

With respect to the completion of tasks with the MyFuture system, the results indicate that the majority of Austrians and Spanish participants believed the system supports them in fulfilling all tasks. They thought it is easy for them to understand the icons. Many of them were able to follow the changes on the screen and they feel comfortable doing all tasks.

Mann-Whitney Tests show a significant difference in the MyFuture post-task questionnaire between Spain and Austria in one item: "I felt comfortable doing the move entry task" ($p=0.01$). Spanish older users are more in agreement with it than Austrian users ($M=17.57$ and 10.15, respectively). There were no gender or age differences in the MyFuture post-task questionnaire.

4 Conclusions

Despite the fact that the older users had some initial difficulties with the on-screen keyboard and the clock used to set the time in the MyFuture application, the results indicate that technology acceptance in general is relatively high. Users generally succeeded in completing the tasks and they were very interested in using the application and saw the value that such a system can have on their lives to support prospective memory. This can be related to the familiar interface of the calendar.

Also, while some users said they still prefer their old paper calendar, many users saw the value of the electronic calendar, especially when they have to cancel an appointment or change the day.

The results that are described above will be incorporated in the next HERMES prototype. The improvements will be made during the following months and will be evaluated in the second user trials in the summer of 2010. These trials will be carried out at the user's home.

Acknowledgments. This work is part of the EU HERMES project (FP7-216709), partially funded by the European Commission in the scope of the 7th ICT Framework. The authors acknowledge the valuable help and contributions from all partners of the project.

References

1. Burdick, D.C., Kwon, S.: Gerotechnology: Research and practice in technology and aging. Springer, New York (2004)
2. Dillon, A.: User acceptance of Information Technology. In: Karwowski, W. (ed.) Encyclopedia of Juman Factors and Ergonomics, Taylor and Francis, London (2001)
3. Nickell, G.S., y Pinto, J.N.: The Computer Attitude Scale. Computers in Human Behavior 2, 301–306 (1986)
4. Buiza, C., Gonzalez, M.F., Etxaniz, A., Urdaneta, E., Yanguas, J., Geven, A., Höller, N., y Tscheligi, M.: Technology Support for Cognitive Decline and Independent Living. Presenting the HERMES Project. Gerontological Society of America Conference (2008)
5. `http://www.fp7-hermes.eu/`
6. Venkatesh, V., Morris, M.G., Davis, F.D., y Davis, G.B.: User Acceptance of Information Technology: Toward a Unified View. MIS Quartely 27, 425–478 (2003)

Agency, Ambience, Assistance: A Framework for Practical AAL

Richard Tynan, G.M.P. O'Hare, and Michael J. O'Grady

Abstract. Ambient Assisted Living (AAL) is a societal imperative in many countries due to their aged demographic profile. Ambient Intelligence (AmI) offers a viable and intuitive framework by which such systems may realised in practice. However, a number of issues must be addressed if AmI is to become sufficiently mature and robust to support AAL. One such issue is the seamless combination of services, based on a diverse range of embedded artifacts, such that AAL systems can be rapidly designed, implemented and deployed. This paper considers, albeit briefly, issues pertaining to the design of an intelligent middleware solution for the provision of AAL services.

1 Introduction

Viewed objectively, Ambient Intelligence (AmI) may be perceived as the latest computing usage paradigm that will change the way people live, work and play forever. Many other paradigms have promised a similar panacea: ubiquitous computing, pervasive computing and wearable computing amongst others. All have fallen far short of what of their proponents and evangelist promise, though time may remedy this. The reasons for this unfilled potential are open to debate, and beyond the scope of this discussion. However, the question arises as to whether AmI will suffer a similar fate. It may well do so; however, it has one critical advantage that the others did not have at their inception. The realisation of AmI as a valid paradigm would, if realised in its entirety, serve as a foundation for the practical deployment of Ambient Assisted Living (AAL) systems. The deployment of such systems is perceived by many as a societal imperative in the coming decade, and this can only occur if many of the issues that hinder the development of practical AmI systems are successfully addressed.

Richard Tynan · G.M.P. O'Hare · Michael J. O'Grady
CLARITY: The Centre for Sensor Web Technologies,
University College Dublin, Blefield, Dublin 4, Ireland
e-mail: {richard.tynan,gregory.ohare,michael.j.ogrady}@ucd.ie

J.C. Augusto et al. (Eds.): ISAmI 2010, AISC 72, pp. 209–212.

2 Ambient Assisted Living

Ambient Assisted Living (AAL) has been conceived as a potential solution to a suite of problems expected to arise in the coming years due the increasingly aged profile of many national populations. It proposes the harnessing of a range of technologies to assist the aged in the course of their everyday activities, for example, in the home, in the workplace or at leisure [AALIANCE(2010)]. If successful, AAL should enable people to live independently for longer. A number of benefits accrue from such an approach, but from a government perspective, cost may the perceived as the most important.

An AAL solution does not demand the availability of an AmI framework. However, if the AmI vision of intuitive and intelligent interaction with a range of embedded computational artefacts was to be achieved, it would indeed form a cornerstone for the deployment of AAL systems. But a number of issues need to be resolved by the research community before this can occur. One of these issues concerns the seamless integration of the diverse range of information sources, for example, embedded sensors, such that activities and contexts can be identified with a reasonable degree of certainty. One approach to this is the development of an open, scalable and extensible middleware solution that provides a uniform view of a range of disparate embedded artefacts such that services based on these artefacts can be rapidly assembled and deployed to meet the diverse contexts and needs in which AAL systems must be deployed.

3 Agent-Based Middleware

In order to bring the components together into a coherent distributed application, a middleware based on intelligent agents has been created called SIXTH. We have adopted an agent based approach to underpin our middleware for a number of reasons. Firstly, it may not be practical to collect all information about the environment in order to reason about the optimal course of action. Agents can be used when decisions must be made with incomplete environmental information. A related benefit to using distributed agents is that in-situ reasoning can limit data traffic and improve the responsiveness of the system. The optimal course of action may not be one action but rather a set of actions that may be carried out sequentially or concurrently. Such planning can facillitate optimal system performance when plans are shared and can allow the agent to reevaluate its current plan.

Current middleware offerings for sensor networks tend to encompass systems that can aid application development to any system between the application and the operating system of the sensor. While this broad definition includes many solutions, there are a number of common goals: to facilitate application development and to provide the capability of deploying a sensor network with little in depth knowledge of WSNs. The SIXTH Middleware is in keeping with these existing high level goals of WSN middleware but also provides additional capabilities not currently present:

modularity, flexibility, reusability, openness, extensibility, universality, multiple abstractions and dynamic reconfigurability.

The middleware architecture that we adopt in this work partitions the network according to the capabilities of the devices. At the top level, devices with the most resources co-operate using our full SIXTH middleware and intelligent agents to deliver system level adaptation. At the intermediate level are devices capable of supporting intelligent agents but not the complete middleware. Finally, devices who operate using a basic embbeded Operating System such as TinyOS or Contiki are in the lowest partition.

One of the primary goals of SIXTH is to facillitate the creation of reusable components capable of intelligent behaviour which can be brought together to form one or more applications. The SIXTH middleware is therefore the collection of components along with their integration framework, which in our case is OSGi. On a marco level this means that the application uses the agent-based middleware components and the integration framework and each component may form part of many applications. Given the uniqueness of the sensor devices, it is important that many applications can coexist on the same sensor infrastructure and we achieve this through the use of a Multi Agent System (MAS). The in-situ agents can be augmented with additional agents as device becomes part of new distributed applications.

When creating applications for devices with sufficient processing capabilities, such as the Sun SPOT or iMote2, SIXTH provides support for the development and deployment of intelligent agents. Specifically, it supports the construction of Agent Factory Micro Edition (AFME) agents [Muldoon et al(2006)Muldoon, O'Hare, Collier, and O'Grady]. AFME is a minimised footprint agent platform that facilitates the execution of intelligent agents on resource constrained devices. Broadly speaking, it conforms to the semantics of the Agent Factory Agent Programming Language [Collier(2001)]. Agents adopt beliefs, which represent their model of the world, and use these beliefs to determine a set of commitments. Commitments represent an intended course of action and will be revised throughout execution as circumstances change.

AFME incorporates functionality for intelligent scheduling, dynamic role adoption, and resource bounded reasoning. These capabilities provide agents with the capability to dynamically alter their behaviour and computational overhead in accordance with quality of service requirements, such as application lifetime requirements. For instance, when power is running low on nodes, agents will reduce their computational overhead by executing their reasoning algorithm less frequently. Additionally, using the resource-bounded reasoning capabilities, agents will drop commitments with a low priority or utility value.

Devices that are unable to support a MAS such as those based on TinyOS or Contiki are incorporated into the overall distributed application through a well defined set of APIs that provide high level abstractions for the application developer. By partitioning the devices of an AAL system into three tiers, we provide functionality in a modular way which can be reused in many different contexts not just AAL.

4 Home Instrumentation and Beyond

In order to evaluate our intelligent middleware in real world context, we have implemented an embedded sensor system which consists of the following elements: passive infra red to detect presence in a room, microphones to sense the ambient sounds in the patient/users home, light and temperature sensors to sample environmental conditions, air quality sesnors, contact and pressure sensors to detect users location on furniture, cameras to monitor user and guests movements, energy and water monitoring to understand appliance activity and usage patterns.

Using this wide range of devices we can develop our reusable intelligent components, deploy them for testing and redeploy them in a different application domain e.g. environmental monitoring. The middleware and intelligent behaviour remains static while the embodiement of the system changes. Reasoning about sampling rate and optimal power usage can be reused with simple reconfiguration of the Quality of Service (QoS) impacts of its decisions to be considered by the agents.

5 Conclusion and Future Work

Ambient Assisted Living is fast becoming a necessity worldwide as the average lifespan of citizens increases. AAL applications are a natural testbed for AmI research, however the development of these systems can be hindered by complexity and limited nature of the devices which the application is based. Our work to date has ficussed on the development of an agent based middleware that can facillitate the deployment of intelligent distributed applications in the AAL space. Crucially, this middleware is built on the guiding principles of modularity, abstraction, reusability and flexibility which will allow the intelligent agents to be reused in other contexts thus shortening the development time and burden of AmI applications.

Acknowledgements. This work is supported by Science Foundation Ireland under grant 07/CE/I1147.

References

[AALIANCE(2010)] AALIANCE, Ambient assisted living roadmap (2010), `http://www.aaliance.eu/public/documents/aaliance-roadmap`

[Collier(2001)] Collier, R.: A framework for the engineering of agent oriented applications. PhD thesis, University College Dublin (2001)

[Muldoon et al(2006)Muldoon, O'Hare, Collier, and O'Grady] Muldoon, C., O'Hare, G., Collier, R., O'Grady, M.: Agent Factory Micro Edition: A Framework for Ambient Applications. In: Alexandrov, V.N., van Albada, G.D., Sloot, P.M.A., Dongarra, J. (eds.) ICCS 2006. LNCS, vol. 3993, pp. 727–734. Springer, Heidelberg (2006)

Automatic Generation of Personal Maps

José Teixeira and Carlos Bento

Abstract. Current collection of data on urban space usage has to rely on paper based surveys that require a huge investment to put in place. New technologies, like positioning systems and handheld devices, provide us with means to build a system that is able to that same data, with less costs and with extended possibilities. For such a system to be successful, it must be aware of the user's activities in the less intrusive way possible. Activities are all associated with the places where they occur. This information associated with the modes of transport used to travel between such places forms the basic components of a person's personal map. In order to build this map, we base ourselves on data collected through the multiple sensors present on a smartphone and use it in conjunction with several artificial intelligence and statistic techniques to build a model capable of inferring activities in a non intrusive way, after a short period of use.

1 Introduction

In our daily lives, we all have a routine to some degree, such as having regular hours for meals, rest or even going to and coming from work. In a general analysis, this routine consists of traveling between the places where we perform the activities we want or need. Information about how we move in a regular basis, and the activities we perform can be extremely valuable in a wide range of sectors and business including, but not limited to, urban and transportation planning.

José Teixeira
Dep. of Informatics Engineering of the University of Coimbra
e-mail: jrrt@student.dei.uc.pt

Carlos Bento
Dep. of Informatics Engineering of the University of Coimbra
e-mail: bento@dei.uc.pt

J.C. Augusto et al. (Eds.): ISAmI 2010, AISC 72, pp. 213–216.
springerlink.com

In this work we propose the automatic creation of what we call *personal maps*, *i.e.* maps containing information about a person's activities, places and transports used using data collected from a *smartphone* device, such as location and sensorial data. We start by analyzing the importance of these maps, and why are we working to make them a reality. Then, we present existing techniques that can aid us in our task, mainly focused on information extraction from location data. Next, we propose the inclusion of new types of sensors and the corresponding extracted information in this kind of analysis, explaining potential use cases and advantages over the current approaches. Finally, we discuss potential "real world" applications for this kind of system.

2 Why Personal Maps?

Information on how a population moves in urban space, and the activities performed in each place is extremely valuable in a wide variety of sectors of society. A bus company can use the travel information to optimize its routes and schedules, a person who wants to open a restaurant can use travel and activity information to determine in what part of the city would it be more profitable, or even what kind of restaurant would be most successful in a specific area, urban planners could more easily analyze where there would be most useful to build a new street or crossroad, bankers would be capable of analyzing where their clients spend the most money in order to adapt the products they offer. These are just a few examples among a virtually infinite number of possibilities that can be explored using this kind of data.

Traditionally, this data is collected by paper and questionnaire based surveys, that are difficult and expensive to develop, distribute and process. However, given the rising availability of devices with great sensing and processing capabilities, that no longer has to be true. We can use these devices to gather accurate and complete data about a person's location and context at any given time. This enables us to extract not only the information we would obtain with questionnaires, but a much more complete vision of one's daily life, with a much lower cost and with a much more immediate availability.

This work is also integrated on the iTEAM project that aims to build and integrated model for transportation and energy consumption based on activity information. Our task is to provide activity information to be used in a micro-simulation model that will return energy consumption and traffic data, allowing for the test of *green policies* in urban and transportation planning.

3 Data Collection and Analysis

In order to collect data from the population, we are going to use *smartphones*. *Smartphones* are devices that take the form of a cell phone but come equipped with much more powerful hardware in terms of processing power, communication interfaces (such as Wi-Fi and Bluetooth) and sensing capabilities (such as GPS, accelerometers and digital compass). Furthermore, with the appearance of devices like the *iPhone* and *android* phones, *smartphones* are becoming more and more ubiquitous, being used independently of profession, ethnicity, religion or social level making them an excellent platform for data collection.

3.1 Location Data

As our focus is obtaining data concerning the user's movement in the urban space and the activities he performs in certain locations, the most important kind of data we want to collect is location data. In order to collect this kind of data, we use the *smarphone*'s GPS receiver. GPS is a global navigation satellite system, i.e. it provides the user with it's location, based on the satellites visible by the receiver, in the form of latitude, longitude and altitude values.

GPS data has been used by many authors to collect many different kinds of information. For our work, we are interested in approaches that allow us to collect information about the users mode of mobility (i.e. if he is walking, driving, cycling, etc.) and the users activities in the places he goes.

Our current prototype implementation of the system is based on the one proposed by Zheng [5] and accomplishes this task by using an approach that combines online inference with offline learning. In the offline learning, previously labeled training data is used to generate an inference model. This model is used online to make determine the mobility mode being used at the moment. For this approach Zheng considers data from sequences of GPS readings done in the same mobility mode such as average, minimum and maximum speed, velocity change rate, stopping rate and heading change rate. While focused on mobility modes, the graph here produced can be used to extract the places where the user performs activities.

We intend to expand this approach using the ideas proposed by Liao [2]. In his model, GPS readings are grouped in situational and navigational activities, and activities are clustered in places important for the user. All this process is done using Relational Markov Networks(RMN) [4] on a relational database, in order to create a graphical model called Conditional Random Field(CRF) [1] which we will use to determine how likely a given mobility mode or activity is, given the actual circumstances.

3.2 Improvements Using Sensor Data

First, satellites need a clear sky, there are places like tunnels or simply indoor environments, where the signal is non-existent thus bringing some problems to the previous approaches. Second, the constant use of a GPS receiver on a battery powered device is still limited to some hours, creating difficulties for an application that needs to capture data on the user on a constant basis.

In order to fight this problem, we propose the incorporation of data collected from accelerometers and digital compass into the existent models. Accelerometer data can greatly improve the detection of the mobility modes and activities (see [3]) by characterizing the user's movements and, when combined with digital compass data, can be used at a lower level to determine the direction where the user is moving, and contribute to reduce the need of a constant GPS signal, extending battery life.

4 Other Challenges

For a system of this kind to work, we need to ask the user to initially insert the needed data and to correct the system in case of errors in the inference. This leads us to an intrusiveness problem. On one hand, we need the user to accurately provide the information making it tempting to ask the user on the spot what is he doing, however, this would most likely lead the user into stop using the application and so, we must equate the possibility of asking the user what he is doing only when he arrives home, or even only when he desires. We need to study how this approach affects the veracity of the information provided, and what can we do to improve it.

5 Conclusion

The information contained in a *personal map* can be very useful in a large number of sectors of society, and the present technology provides us with the means to generate it automatically. In order to do this, we use models based on location data and intend to improve them both by combining existing efforts and by adding extra data collected from the accelerometer and digital compass present on a smartphone.

References

1. Lafferty, J., McCallum, A., Pereira, F.: Conditional random fields: Probabilistic models for segmenting and labeling sequence data. In: Proceedings of Eighteenth International Conference on Machine Learning (2001)
2. Lin, L.: Location-based Activity Modeling. PhD thesis, University of Washington (2006)
3. Ravi, N., Dandekar, N., Mysore, P., Littman, M.L.: Activity recognition from accelerometer data (2005)
4. Taskar, B., Abbeel, P., Koller, D.: Discriminative probabilistic models for relational data. In: Proceeedings of Conference on Uncertainty in Artificial Intelligence, Edmonton (2002)
5. Zheng, Y., Li, Q., Chen, Y.: Understanding mobility based on GPS data (2008)

Modeling and Simulating Ambient Assisted Living Environments – A Case Study

Tim Laue and Christoph Stahl

Abstract. Modeling the layout of an apartment – including different furniture configurations and several moving elements – is a non-trivial task, especially if accessibility has to be assured e. g. for wheelchairs. In this paper, we describe a two-staged example for such a task by connecting two different applications: In the first step, the *YAMAMOTO* toolkit is used to efficiently model the building structure and to plan the furnishing of the environment in 3-D. Afterwards, the desired configuration becomes exported to *SimRobot*, a robot simulator based on rigid body dynamics. Thereby, a realistic evaluation of the physical configuration becomes possible, e. g. by interactively driving an electrical wheelchair through the environment.

1 Introduction

Imagine a person named Mario who depends on an electrical wheelchair and lives in an apartment providing ambient assistance. Mario sits in his wheelchair in the bedroom and tells the system: *"I would like to drink a coffee"*. Thereupon, the wheelchair starts to drive autonomously to the kitchen, bypassing any furniture blocking the way. The doors open automatically and whenever a new room is entered, the light becomes switched on. During the ride, the kitchenette adjusts to a height which suits the height of the wheelchair. When Mario arrives in the kitchen, the wheelchair parks in a position which enables Mario to easily access the cupboard's content.

This short scenario describing a rather simple everyday task involves a number of different components which together form a complex system. Whenever a change of the environment – e. g. the addition of new furni-

Tim Laue · Christoph Stahl
German Research Center for Artificial Intelligence (DFKI)
e-mail: {Tim.Laue,Christoph.Stahl}@dfki.de

J.C. Augusto et al. (Eds.): ISAmI 2010, AISC 72, pp. 217–220.
springerlink.com

ture or a change of the wheelchair model – occurs, the interplay of all parts is not guaranteed anymore. Therefore, careful planning and probably time-consuming evaluation are necessary in advance to avoid any fundamental problems. The contribution of this paper is a first step towards an efficient process for modeling and evaluating new configurations in such a scenario.

This paper is organized as follows: Section 2 describes the necessary preliminary works in the areas of Ambient Assisted Living, rehabilitation robotics, robot simulation, and modeling. The applied process for modeling and evaluation is described in Sect. 3, followed by the conclusion in Sect. 4.

2 Preliminary Works

The Bremen Ambient Assisted Living Lab (BAALL) [1][`http://baall.net`] is an apartment of 60 m^2 suitable for the elderly and people with physical or cognitive impairments. It is situated at the Bremen site of the DFKI. The goal of the project is to investigate how the living environments of seniors to-be can be instrumented with infrastructures that allow incremental upgrades with user assistance systems. In order to allow a wheelchair-dependent user to stay together with their non-impaired partner in the same living environment, user-adaptable furniture has been installed (cf. Fig. 1).

The Intelligent Wheelchair Rolland is a mobility assistant that is able to operate within the BAALL. It is based on the commercial power-wheelchair *Xeno* by Otto Bock Mobility Solutions, see Fig. 1a. The wheelchair is upgraded with two laser range finders, wheel encoders, and an onboard computer. Various driving assistant modes for the wheelchair Rolland are being developed and evaluated to compensate for diminishing physical and cognitive faculties, e. g. the *driving assistant* avoids obstacles and facilitates passing through a door, and the *navigation assistant* guides along a route or drives autonomously.

a)  b) 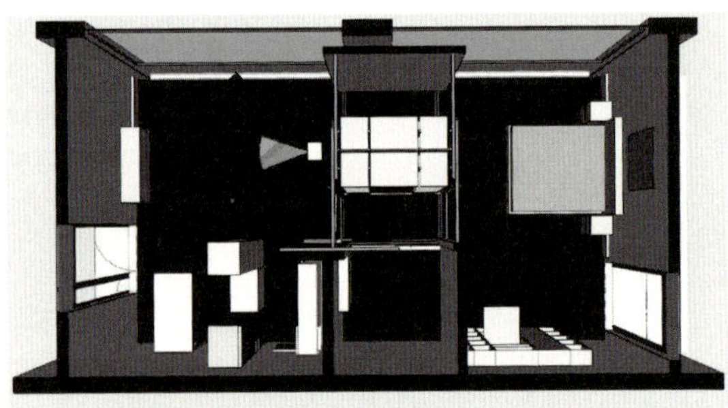

Fig. 1 a) The wheelchair Rolland in the BAALL kitchen. The kitchenette and one cupboard have been lowered automatically. b) The BAALL modeled in *YAMAMOTO*.

SimRobot is a robot simulator which is able to simulate arbitrary user-defined robots in three-dimensional space [2]. To allow an extensive flexibility in building accurate models, a variety of different generic bodies, sensors, and actuators has been implemented. One main application has always been the simulation of autonomous wheelchairs but up to now, a variety of legged and wheeled robots has been successfully simulated by SimRobot.

YAMAMOTO (*Y*et *A*nother *MA*p *MO*deling *TO*olkit) is a graphical editor that we have developed for the geometric modeling of co-located buildings and intelligent environments in 3-D [4]. The toolkit has been designed to efficiently model the building's structure from floor plans. Additionally, the furnishing of the environment can be used through pre-defined shapes (cf. Fig. 1). Furthermore, the toolkit supports the planning and design of intelligent environments by providing various sensor and actuator elements [3].

3 Simulation of Physical Assistance in BAALL

During the course of the planning of BAALL, one practical design problem was rised concering the furnishing of the living space with shelves and kitchen appliances: How can the storage space be maximized without compromising the maneuverability of the wheelchair? We aimed towards a physical simulation of the wheelchair within the planned BAALL environment. We have already started to model the BAALL environment with *YAMAMOTO* for visualization and documentation purposes. The model comprises the building structure, the sliding doors, and all furnishing items. In order to simulate the dynamic aspect of the sliding doors, we have extended the export module of *YAMAMOTO* to automatically create RoSiML (Robot Simulation Markup Language, used by SimRobot) code that specifies a fully functional physical model of the motorized doors. It is important to say that without a semantic knowledge layer of the parametric objects in *YAMAMOTO*, this would not have been possible. If the sliding doors were only represented as geometric primitives (boxes, triangles) without further type information, as usual in 3-D modeling tools, the export module would not know how to create the necessary physical elements. Finally, we have added a physical model of the wheelchair to the SimRobot scene. The model provides a realistic driving behavior and can be interactively controlled by the user. The simulated BAALL environment allows the evaluation of the wheelchair's mobility in bottlenecks, such as the kitchen, cf. Fig. 2. We made the experience that it quite difficult to steer the wheelchair into the kitchen – we always needed several trials and often got stuck, despite having an allocentric perspective. In addition, the previously determined wheelchair configurations allowing a convenient access to different kitchen elements also became harder – in some cases not at all – to reach. Based on this result, we gave the advice to rethink the planned kitchen configuration.

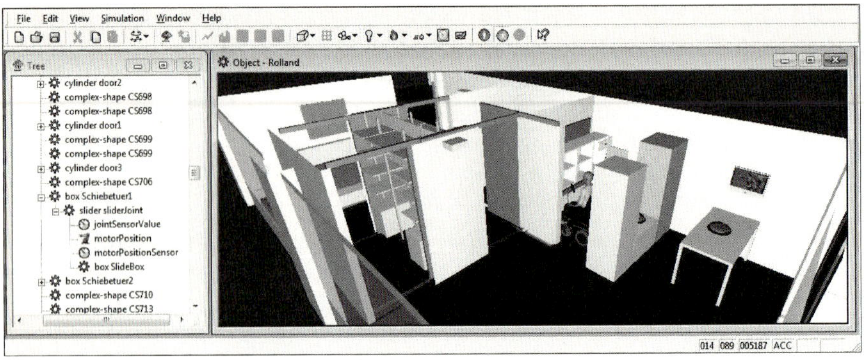

Fig. 2 Assessing the maneuverability of the wheelchair in the dynamic simulation

4 Conclusion and Future Works

In this paper, we showed how the design of a complex environment such as an Ambient Assisted Living lab interacting with a mobility assistant can be modeled and evaluated using powerful, already existing tools. We have created a virtual model of an ambient assisted living lab that includes physical elements and applied this simulation to assess the maneuverability of a wheelchair in a planned kitchen configuration. However, in this first case study, only the maneuverability of the wheelchair in case of manual steering has been tested. For the future, we plan to extend the evaluation in the simulation from the manual control mode to assisted driving modes provided by the real wheelchair. The virtual environment would be ideal to test how the assistant copes with the kitchen situation as well as to reduce the need for testing in the lab. In case of new environments that have not yet been built, such a virtual simulation environment will become absolutely necessary to test the compliance of the driving assistant with the proposed furniture.

References

1. Krieg-Brückner, B., Gersdorf, B., Döhle, M., Schill, K.: Technik für Senioren in spe im Bremen Ambient Assisted Living Lab. In: 2. Deutscher AAL-Kongress 2009, Berlin-Offenbach, Germany, VDE-Verlag (2009)
2. Laue, T., Spiess, K., Röfer, T.: SimRobot - A General Physical Robot Simulator and Its Application in RoboCup. In: Bredenfeld, A., Jacoff, A., Noda, I., Takahashi, Y. (eds.) RoboCup 2005. LNCS (LNAI), vol. 4020, pp. 173–183. Springer, Heidelberg (2006)
3. Stahl, C.: Spatial Modeling of Activity and User Assistance in Instrumented Environments. PhD thesis, Dept. of Computer Science, Saarland University (2009)
4. Stahl, C., Haupert, J.: Taking location modelling to new levels: A map modelling toolkit for intelligent environments. In: Hazas, M., Krumm, J., Strang, T. (eds.) LoCA 2006. LNCS, vol. 3987, pp. 74–85. Springer, Heidelberg (2006)

Methodology for the Integration of Information, Communication and Automation Technologies in Housing Rehabilitation

Sara Eloy, Renato Nunes, Isabel Plácido, and José Pinto Duarte

Abstract. This paper addresses the integration of Information, Communication and Automation Technologies (ICAT) in the dwelling space in order to meet the user's needs. We describe an ongoing PhD research which deals with the rehabilitation of the existing housing stock in order to fulfil the new needs of dwellers in the current Information Society as well as the impact of the integration of ICAT in dwellings. For this purpose we establish a rehabilitation methodology which enables architects to fulfil the client's needs and requirements in home automation, from the beginning of the architectural process. This will enable the creation of a compatible and properly integrated ICAT infrastructure in homes, paving the way for ambient intelligence and promoting sustainable strategies at the environmental and social levels.

1 Introduction

With this research we are dealing with two fundamental problems: the urgency of rehabilitating the existing housing stock and the urgency of integrating ICAT in homes.

Much of the existing housing stock in Lisbon needs rehabilitation due to both constructive pathologies as well as to emergent conflicts in the use of spaces that result from changes in current lifestyles. New approaches to the design of domestic

Sara Eloy · Renato Nunes · José Pinto Duarte
Technical University Lisbon, Portgal
e-mail: sara.eloy@iscte.pt, renato.nunes@ist.utl.pt

Sara Eloy · Isabel Plácido
National Laboratory for Civil Engineering, Lisbon, Portgal
e-mail: iplacido@lnec.pt, jduarte@fa.utl.pt

J.C. Augusto et al. (Eds.): ISAmI 2010, AISC 72, pp. 221–228.
springerlink.com

spaces will answer theses demands and meet the need for accommodating activities such as telework and telehealth, in response to the growing demand of information access, comfort and safety in homes. The first research goal is to identify how the use of technology influences the ways of living and creates new dwelling requirements, and how this impacts the spatial and functional organization of dwellings. This work complements Pedro's (2000) and Duarte's (2001) frameworks and Oliveira's research (2000) to incorporate new dwelling modes, new compositions of domestic groups, and ICAT-related demands. The integration of ICAT in housing is a priority in the quest for improved housing quality, comfort and adaptability. ICAT also promotes ecological sustainability as well as social sustainability by promoting social inclusion of citizens and supporting the elderly and people with reduced mobility. Besides the advantages of domotics in assisting the elderly and disabled persons, these technologies assist all people in their daily tasks, making everyone's life more comfortable and safe. Considering the importance of providing ICAT in houses, one of the fundamental goals of the ongoing research is the definition of adequate sets of ICAT to incorporate in the dwelling spaces according to family profiles. The provision of such technologies aims at guaranteeing sustainability and the social integration of citizens, while adapting them to each household, according to its present and future needs. Such set of ICAT would be applicable to the dwelling as well as the building, not only in new construction but also in the rehabilitation of the existing residential stock. The last goal of this research is to define design guidelines, as well as a methodology, to support architects in the adaptation of existing dwellings and in the incorporation of ICAT while allowing them to balance new dwelling trends with sustainable requirements and economic feasibility.

This paper will focus on both the definition of adequate sets of ICAT according to family profiles as well as the proposed methodology for integrating them in housing rehabilitation.

2 ICAT as Ecological and Social Sustainability Strategies

Home automation systems can have a very important role in contemporary dwellings because of the benefits they bring to residents both regarding ecological and social aspects. Domotics represents a means to manage household energy use and to complement, not to substitute, passive solutions. A well insulated building is the most effective element of an efficient heating, ventilating, and air conditioning (HVAC) system. Therefore, automated domotic controls for HVAC systems represent a complement to the passive strategy. The often called "Smart House" will only be smart if, additionally to automation controls, it is designed and built in a sustainable way. This means environmentally-conscious design techniques in the field of architecture using efficient insulation, earth friendly building materials, energy and water conservation, solar energy, wind-power, among other strategies. However, when actuating over existing buildings the achievement of efficient passive solutions often fails due to morphological and construction restrains. In such cases, where passive solutions prove to be insufficient, active systems are required to increase comfort standards. Better control systems for heating and cooling management are fundamental to achieve that goal and promote energy efficiency. There are several domestic automation solutions that can be used for cutting energy consumption, including presence detection devices, solar control systems (intelligent control of blinds and other protective shades that can be lowered, raised and tilted according to the detected location of the sun), and computer controlled ventilation by motorized opening of doors and windows. In addition to promoting ecological sustainability, ICAT helps improving the lives of inhabitants as it can anticipate and respond to resident's demands, thereby promoting social sustainability. Smart houses can also serve the elderly and people with reduced mobility, as well as other groups of inhabitants, assuring their comfort and contributing to their safety, by allowing them to surpass obstacles and supporting the execution of their daily tasks. Assistive technologies can be designed to evolve as the resident requires more assistance around the home and helps them living more independently.

3 Defining the Methodology

Ongoing research intends to define a rehabilitation methodology for Lisbon's existing housing stock so that it responds to new technology requirements and new ways of living. Our work started with the analysis of dwellers demands as well as with the development of a knowledge base regarding the set of existing ICAT for homes, which will be taken into account during the application of the rehabilitation methodology. Then, we raised a hypothesis for such a methodology based on the conceptual schema of the design process proposed by Duarte (2001) for the mass customization of housing. It was hypothesized that a rehabilitation methodology should encompass four steps, as shown in Fig. 1.

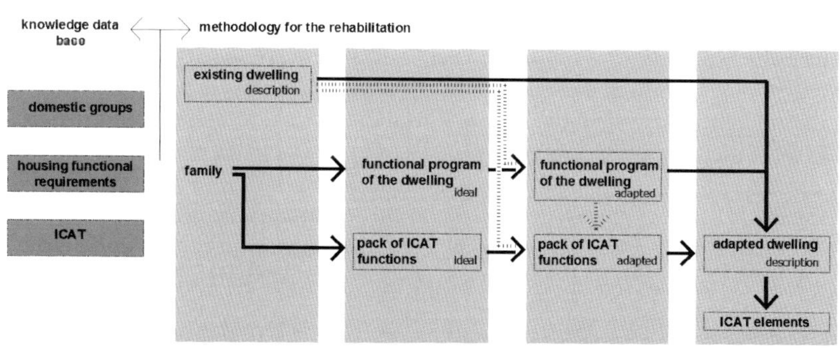

Fig. 1 Basic steps in the envisioned rehabilitation methodology

The first step consists in gathering the data needed for the rehabilitation process: the profile of the future dwellers and a description of the existing dwelling. In the second step the family profile is used to elaborate the ideal functional program of the dwelling as well as the ideal pack of ICAT functions. The ideal pack of

ICAT functions is a description of an ideal set of systems and functions that respond to a given family profile. This pack of ICAT is not bound to any existing morphological or construction restrains. Similarly, the ideal functional program is a description of the ideal housing solution for that family regarding spaces, rooms and topology relations between them. In the third step, the existing dwelling and both the ideal functional program and the ideal pack of ICAT, are used to derive a description of a compromised or adapted solution based on the existing dwelling. Since the solution is conditioned by the existing morphological structure, a third step is necessary to transform the description of an ideal solution into the description of an adapted one. Finally, from the description of the adapted dwelling is obtained the layout of a design solution for that family in that dwelling, which includes the ICAT components needed in the dwelling. For the purpose of incorporating ICAT in dwellings two complementary methods were established. First, the introduction of ICAT in dwellings changes some aspects of living (e.g. home cinema, telework) which were taken into account in the definition of today's houses functions. Second, the introduction of ICAT in dwellings operates physical changes in the house prompted by the need for accommodating cabling infrastructures and terminal equipment.

4 Packs of ICAT Functions

Smart homes should address the inhabitant's demands and needs. However there is frequently a gap between the implemented technology and user's real needs. This gap occurs because the implementation of home automation is usually undertaken without previous knowledge of end-users (e.g. in multi-family residential buildings) or without understanding the exact needs of residents (). It is important that every stakeholder within the building process, in particular the architects and clients, have enough knowledge of ICAT so that the pack of technologies chosen can be used in a useful way by the end-user.

To achieve the definition of the ideal pack of ICAT functions we considered several types of households and, based on their characteristics and anticipated needs and activities within the home, we establish a set of essential domotic functions. This method follows previous works on: i) user-centered approach by Meyer and Schulze (1996) on uptake of technology based on household characteristics; Bierhoff and Berlo (2007), on technology preferences based on lifestyles; Becker and Van de Goor (2002) on technology generation; ii) centered approach on activities which take place at home by Mateas et al. (1996), iii) hierarchical classes of smart home by Aldrich (2003).

For each family profile we defined three levels of ICAT packs that can be chosen according to the family's economic capacity. The minimal proposed level of ICAT is recommended to every new or rehabilitated house or dwelling. This level includes wired infra-structures, basic security systems and access to ICT, and the exact definition of these features varies slightly according to different family's profiles. This minimum level contains all cable infra-structure needed to allow upgrading the system up to the optimal level. The minimum level is suitable for low budgets households and property developers who want to make a small investment

Table 1 *Assignment of ICAT functions according to family profile*

	Couple with children				Couple with-out children				Monoparen-tal families				Co-habita-tion		1 house-hold		
Home automation functions	With small children	With grow up children	With elderly or disable person	With other persons	May have children in future	Without children as a option	Children already left home	Elderly or disable person	With small children	With grow up children	With elderly or disable person	With other persons	Any age	With elderly or disable person	Young	Middle-age	Elderly or disable person
Safety / Security																	
Access control	L2	L2	L2	L2	L2	L2	L2	L1	L2	L2	L1	L2	L2	L1	L1	L1	L1
Intrusion detection	L1	L1	L1	L1	L1	L1	L1	L1	L1	L1	L1	L1	L1	L1	L1	L1	L1
Surveillance systems	L3	L3	L3	L3	L3	L3	L3	L3	L3	L3	L3	L3	L3	L3	L3	L3	L3
Smoke detection	L1	L1	L1	L1	L1	L1	L1	L1	L1	L1	L1	L1	L1	L1	L1	L1	L1
Gas leak detection	L1	L1	L1	L1	L1	L1	L1	L1	L1	L1	L1	L1	L1	L1	L1	L1	L1
Water leak detection	L2	L2	L1	L2	L2	L2	L2	L1	L2	L2	L1	L2	L2	L1	L1	L1	L1
Lack of energy detect.	L3	L3	L2	L3	L3	L3	L3	L1	L3	L3	L2	L3	L3	L2	L3	L3	L1
Personal safety alarm	L3	L3	L2	L3	L3	L3	L3	L1	L2	L3	L1	L3	L3	L1	L3	L2	L1
Comfort																	
HVAC control	L2	L2	L2	L2	L2	L2	L2	L1	L2	L2	L2	L2	L2	L2	L2	L2	L1
Lightning control	L3	L3	L2	L3	L3	L3	L2	L1	L3	L3	L2	L3	L3	L2	L3	L3	L1
Blinds control	L2	L2	L1	L2	L3	L2	L1	L1	L2	L2	L1	L2	L2	L1	L3	L2	L1
Energy management																	
Energy distribution	L2	L2	L2	L2	L2	L2	L2	L2	L2	L2	L2	L2	L2	L2	L2	L2	L2
Domestic equipment	L2	L2	L1	L2	L2	L2	L1	L1	L1	L1	L1	L1	L2	L1	L2	L2	L1
Information / communication																	
Alarms for exterior	L1	L1	L1	L1	L1	L1	L1	L1	L1	L1	L1	L1	L1	L1	L1	L1	L1
Local alarms	L1	L1	L1	L1	L1	L1	L1	L1	L1	L1	L1	L1	L1	L1	L1	L1	L1
Remote control / phone	L1	L1	L1	L1	L1	L1	L1	L1	L1	L1	L1	L1	L1	L1	L1	L1	L1
Remote control / net	L2	L2	L2	L2	L2	L2	L2	L3	L2	L2	L2	L2	L2	L2	L2	L2	L3
Wired bus network	L1	L1	L1	L1	L1	L1	L1	L1	L1	L1	L1	L1	L1	L1	L1	L1	L1
Telehealth	L3	L3	L2	L3	L3	L3	L3	L2	L3	L3	L2	L3	L3	L2	L3	L3	L1
Telecare	L3	L3	L2	L3	L3	L3	L3	L1	L3	L3	L1	L3	L3	L1	L3	L3	L1
Entertainment																	
Home cinema	L2	L2	L2	L2	L2	L2	L2	L2	L2	L2	L2	L2	L2	L2	L2	L2	L2
TV/Net access,distrib.	L1	L1	L1	L1	L1	L1	L1	L1	L1	L1	L1	L1	L1	L1	L1	L1	L1
Audio distribution	L3	L3	L3	L3	L3	L3	L3	L3	L3	L3	L3	L3	L3	L3	L3	L3	L3
Integration levels	Minimum level (L1)						Medium level (L2)						Optimum level (L3)				

without compromising the upgradability of the system. The medium level is an upgrade of the lower level and it requires a greater financial investment. This level includes automation technologies that go beyond the basic setup. The optimum level of ICAT integrates a complete set of technologies for the prospective family needs, without becoming unreasonable.

A simplified version of the assignment of ICAT functions according to family profiles is shown in Table 1 considering the above mentioned levels of integration. We classified families based on their kinship, age and number. We added a class for elderly and disabled people to each one of the previous classes, so that a household with a disabled person incorporates assistive technologies. The pack of assistive technologies is not shown in this paper.

In addition to the pack of ICAT we defined construction strategies to incorporate domotic appliances as well as wire and wireless infra-structures in a rehabilitation work. These strategies take into account that by intervening in an existing building, rehabilitation requires one to manage different construction techniques and cabling demands. Simultaneously, one needs to ensure space flexibility considering both present and future needs. The proposed methodology aims at defining the adequate set of ICAT functions and their integration following the four steps described earlier and illustrated in Fig. 1.

5 Methodology Application – An Example

Adapting the dwelling to the family's needs according to the new functional program and ICAT pack, involves a series of steps. We have defined three different rehabilitation strategies accordingly to the spatial transformation of the dwelling. In the current research, we are using what we call a *transformation grammar* which will allow us to transform an existing design into a new one that matches given requirements. Transformation grammars drift from shape grammars whose

Table 2 Steps to follow in the adaption of a dwelling

steps	Spatial transformation		Incorporation of ICAT elements
0	Locate social and private areas	7	Define consumers: 7.1 sensors (water, motion, temperature, smoke, light); 7.2 controlled lights and power sockets; 7.3 controlled motor blinds; 7.4 electric valves, 7.5 electric oven, 7.6 ...
1	If functional program has 2 or more bedrooms • define private area		
2	If functional program has 2 or more bedrooms • define social area	8	Define operating points: 8.1 control panels; 8.2 push buttons
3	Resolve circulation (social and private area)	9	Define network (diagram)
4	Define service area	10	Create a consumer list
5	Define storage spaces	11	Create a list of operating points
6	Resolve circulation between different circulation spaces		

investigation started more than thirty years ago with Stiny and Gips (1972). Shape grammars are algorithmic systems with a set of rules that apply to shapes in a step-by-step way to generate a language of designs.

The list of steps to follow in the rehabilitation according to one of the rehabilitation strategies is systematized in Table 2. Due to space limitations, in this paper we only show a sample of the spatial transformation and the integration of ICAT in order to illustrate how the grammar works and can be applied – see Fig. 2.

Fig. 2 Partially derivation of one rehabilitation transformation

6 Conclusion

The incorporation of intelligent features adds value to homes and enhances the inhabitants' quality of live, while contributing to better housing management. Home automation technologies increases comfort in houses and helps inhabitants in daily domestic tasks promoting independency. This fact is of major importance to elderly and people with reduced mobility or other disabilities. In addition, the rehabilitation of Lisbon's existing housing stock is a priority in the political agenda and must fulfill new space-use requirements demanded by the new information age lifestyles as well as meet ecological, social, and economical sustainability. In our research we propose strategies that permit to achieve an adequate balance between the tradeoffs involved in the rehabilitation of existing dwellings, the compliance with the new demands on housing functions and the integration of ICAT. The proposed methodology aims at supporting the adaptation of existing dwellings and the incorporation of ICAT according to user's real needs, promoting the development of ambient intelligence.

Currently we have already done some successfully experiments in which we were able to use grammar rules to transform dwellings both on functional aspects as well as introducing ICAT according to family profiles. Future work will be concerned with completing the transformation grammar and fine-tuning the relationship between the grammar rules and the integration of ICAT.

Acknowledgments. The research described in this paper project was funded by the Portuguese Science Foundation (FCT) with grant SFRH / BD / 18225 / 2004.

References

1. Aldrich, F.K.: Smart Homes: Past, Present and Future. In: Harper, R. (ed.) Inside the smart home, Springer, London (2003)
2. Henk, B., Van de Goor, A.-G.: Technology Generations in the Netherlands A Sociological Analysis. Shaker Publishers, Maastricht (2001)
3. Bierhoff, I., van Berlo, A.: Smart home Technology: on the move to ambient and invisible intelligence. In: Keijer, U., Sandstrom, G. (eds.) Smart Homes & User Values, pp. 45–67. The Urban International Press, Great Britain (2007)
4. Duarte, J.P.: Customizing Mass Housing: A Discursive Grammar for Siza's Malagueira houses. Thesis submitted to the Department of Architecture in Partial Fulfillment of the Requirements for the Degree of Doctor of Philosophy in Design and Computation. Massachusetts Institute of Technology, Cambridge, E.U.A (2001)
5. Juez, C.S.: La Realidad de un Sueño. Vivenda domotizada para personas com discapacidad. Madrid, Asociación ADAMAR, Asociación DATO, DULCIT S.L (2007)
6. INTEGER (s.d) INTEGER Intelligent & Green. Tomorrow's Housing Today. Good Practice Guide
7. Mateas, M., Salvador, T., Scholtz, J., Sorensen, D.: Engineering Ethnography in the Home. In: CHI 1996 Electronic Proceedings (1996)
8. Meyer, S., Schulze, E.: The Smart Home in the 1990s. Acceptance and Future Usage in Private Households in Europe. In: The Smart Home: Research Perspectives, The European Media Technology and Everyday Life Network (EMTEL), Working Paper No. 1, University of Sussex, Brighton (1996)
9. Oliveira, F.C.A.N.: Os Novos Lugares do Habitar e as formas de apropriação contemporâneas. Lisboa, FAUTL. Master thesis – Arquitectura da Habitação (2000)
10. Pedro, J.B.: Definição e avaliação da qualidade arquitectónica residencial. Lisboa, Faculdade de Arquitectura da Universidade do Porto. Thesis submitted for the Degree of Doctor of Architecture (2000)
11. Stiny, G., Gips, J.: Shape Grammars and the Generative Specification of Painting and Sculpture. In: Freiman, C.V. (ed.) Information Processing, vol. 71, pp. 1460–1465. North Holland, Amsterdam (1972)

Why Traders Need Ambient Intelligence

Javier Martínez Fernández, Juan Carlos Augusto,
Ralf Seepold, and Natividad Martínez Madrid

Abstract. Trading is widely recognized as a stressful profession. Taking decisions under stress negatively affects both, the finances and the health of the trader. Providing support to increase the awareness of traders on their real levels of stress could significantly improve their performance and well-being. This paper surveys technology available that can be considered to provide such level of support. We explore Ambient Intelligence solutions, which can provide support for two cases: a single trader and a network of traders.

Keywords: Ambient Intelligence, Trading, Stress, Decision-Making.

1 Introduction

The world of Stock Markets and the world of Ambient Intelligence seem dissociated. This paper shows that Ambient Intelligence can be very relevant to the world of trading by supporting individual traders or clusters of them. Trading is usually stressful and more even so in these days of widespread financial crisis which makes decision-making in the stock market harder and harder.

Research on how stress influences decision-making is not new ([1], [2]) Traders depends on good judgment exercised during their work in the financial market.

Javier Martínez Fernández · Natividad Martínez Madrid
University Carlos III of Madrid, Spain
e-mail: {javier.martinez,natividad.martinez}@uc3m.es

Juan Carlos Augusto
University of Ulster, United Kingdom
e-mail: jc.augusto@ulster.ac.uk

Ralf Seepold
University of Applied Sciences Konstanz, Germany
e-mail: ralf.seepold@htwg-konstanz.de

J.C. Augusto et al. (Eds.): ISAmI 2010, AISC 72, pp. 229–236.
springerlink.com

Bad decision-making by a trader under stress can bring catastrophic consequences, not only on her/his finances, also on the psychological health of the trader [3]. How could the trader know when he/she is becoming stressed? Ambient Intelligence as an emerging area focused on building *digital environments that proactively, but sensibly, support people in their daily lives* [4] is relevant to the problem. We argue in this paper that one potential benefit of this technology is to support traders in their daily decision-making so that they become aware of their real levels of stress and can operate in the market in a secure way.

In the next section, we give a brief account of how traders could measure levels of stress with current technology. Section 3 explains why the current technological offer is not enough to provide satisfactory support to traders. The difference between the activity of an Ambient Intelligent assisted financial operator and a traditional market without that support is illustrated in section 4. Section 5 explains how such Ambient Intelligent support can be developed to achieve a positive synergy between the worlds of Ambient Intelligence and Stock Markets.

2 Measuring the Stress State of the Trader

Decision-making is somehow safe as long as the environment conditions do not affect the trader. Medicines like Bitalin, Adderall, Vicodin and Oxycontin are currently widely used in Wall Street to avoid panic moments by controlling the emotions of traders [5]. Bad decision-making can cascade in a negative way. If a trader loses in one decision that can negatively affect the next decision. Detecting the moment in which the trader enters in stress to avoid possible errors in the operations is therefore very important. Right now, there is no way to know if a trader is in stress because the trader himself does not realize of that until it is too late. Hence it is useful to consider whether we can improve the state of the art through technology. There are several studies of human behaviour in real time decision making, about how stress can influence this behaviour, and what biometric variables change in our body. For example, J. M. J. Coates and Herbert et al. [6] have found out that cortisol level change with stress. Other researches ([7], [8], [9]) showed there is a relation between heart rate variability derived from the electrocardiogram (ECG), blood pressure, and stress. Work reported in [10] describes the relationship between the changes in heart rate, blood pressure, skin temperature, and muscle tension in stressful moments. Skin conductance [11], the breathing rate [12], and the brain waves [13], are related to stress too. Based on these studies we can conclude that there are seven main variables that change in stress situations: heart rate, blood pressure, breathing rate, brain waves, muscle tension, skin conductance and temperature. There are some e-health biometric devices (see Section 5) suitable to measure these variables of the trader in real time (unlike drugs, e.g., cortisol), so if we can measure these variables in real time, we have gained an understanding of the trader's stress that can be fed back to her/him. This trader will have the advantage knowing his/her real stress state in real time. We call this an Ambient Intelligent Supported Trader.

3 Ambient Intelligence Traders Network: A Real Time Stress Feeling of the Market

One of the most important problems in trading is the difficulty to realistically gather in real time an indicator of the level of stress experienced by the market, i.e. an immediate measurement of the collective level of stress that traders operating are experiencing. The collective reaction to news during the session, for example, could cause high stress state (even panic) and risky decisions. According to Lori [14] the movements of the stocks markets are very connected with the emotional state of the collective emerging opinion and reaction of traders that are operating in the market itself. But, how to gather stress-related information from the market in real time? Until now, that information is not available to traders. Currently, the closer data available is the sentiment of the market. The sentiment of the market tries to predict the behaviour of the market (if it will go up or down) according to statistics based on the opinion of the traders about the market behaviour for the following week. M. Baker et al [15] show the influence of these statistics in the market and how the sentiment is measured, but the stress never is measured. The most used statistic is the Investors Intelligence Sentiment Index (Advisors Sentiment, 2010, http://www.investorsintelligence.com/) but all the statistics have the following problems:

- It always arrives late. The statistics always are about data of the past, in this case, about the feeling of the trader in the moment in which he/she was enquired. It is helpful to understand the past, but the financial markets are the clearest example of business on real time, and a single piece of news can fundamentally change traders' perception very quickly.
- How reliable is a statistic of feeling in the financial market? It is important to understand that the financial market is the world of money, and this fact influences very much the honesty of people's replies.

Ambient Intelligence appears as a possible solution to gather that important information in real-time. A possible way ahead is the deployment of a "Network of Ambient Intelligence Traders" in an investment company according to the following principles:

- The traders start the session connected to devices that can measure stress in real time.
- Each trader sends his/her data in an anonymous way.
- Each trader could see his own stress as well as the stress of the network in a chart. In this way, the trader knows whether he/she is conditioned by stress or if this is a delicate moment in the market.
- One supervisor could see the stress of each user and the stress of the network. In this way, if one user is in an abnormal state and continues operating the supervisor could advertise to him. Besides, with the stress network, the supervisor could detect delicate moments in the market.

Figure1 shows the described schema of work where all traders send the stress data to the supervisor and Figure 2 shows the individual stress data of the traders in pink and sentiment of the market in blue (gathering all data of the traders).

Fig. 1 Work Schema

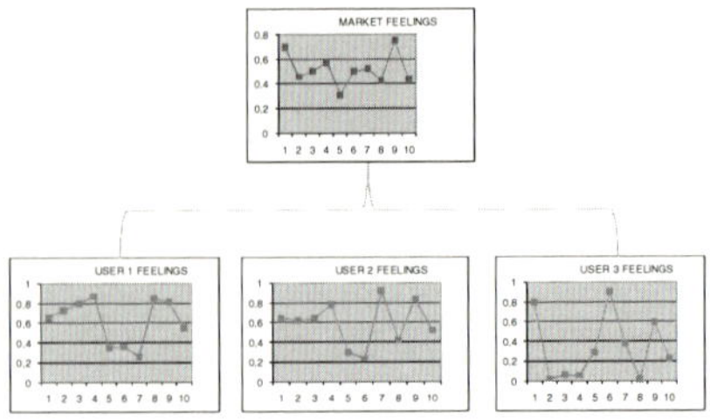

Fig. 2 Charts of Stress

If Ambient Intelligence provides us a way to find the real stress state of one trader in real time, then, if we merge networks of such traders, we can have an indicator of the real time stress feeling of the market. With this schema of work, many investment companies could avoid "Black Days" where high stress moments cause panic and risky decisions. The next section illustrates the difference Ambient Intelligence support can make during one trading session.

4 Ambient Intelligence Trader vs Traditional Trader: A Workday

A trader's workday is always difficult; each day is as a chess game with different pieces. The following tables illustrate the difficulties associated with market trading and the potential role of Ambient Intelligence. The scenario is based on an IBEX-35 (Spanish stock market index of reference). Table 1 shows how without Ambient Intelligence support, trader A, is conditioned during the first moments of the sessions by his/her emotional state, but A is not aware of it so A's trading process from this moment (9:00) onwards would be at risk and decisions A makes could be wrong. Trader B must be making decisions by (13:00) which are based on his intuition and on weekly statistics, because he/she does not know the real time situation of the traders at that moment. In Table 2, at the beginning of the session (9:00) trader C notices his/her state is abnormal, hence C decides to stop the trading process. The Ambient Intelligence support allows traders to take necessary actions to change his abnormal state. Trader D makes decisions with the real time perception of the market in mind in the second important moment of the session (13:00) and hence D's decisions are better.

Everybody has bad days, and Ambient Intelligence support does not guarantee to win more money, but it supports better informed decision-making and, in the financial market, this is the best help that a trader could have. Crucially it can make the traders self-conscious of their emotional state as well as of the market emotional status as a whole. The remaining question is: can Ambient Intelligence

Table 1 Trader A's bad day vs trader B's normal day, both without AMI support

Time	Trader A Bad Day	Trader B Normal Day
9:00 -9:05	Pre-Open Market. A manages many contracts. A is stressed. Besides, A had an argument with his/her partner.	Pre-Open Market. B manages many contracts, and B is stressed.
9:05-13:00	Opened Market. The session is quiet, but A is unconsciously conditioned by the argument and A starts to fail in the trading process.	Opened Market. The session is quiet, and B has a normal trading process.
13:00-13:05	An important data is published. Stress is high but the report requires action. Decisions are taken under bad emotional conditions for A.	An important data is published. B does not know the perception of the people so B takes decisions with the information of the statistics with previous data.
13:05-15:30	A had failed before and is being conditioned by the argument plus previous bad decisions. As A feels the needs to do it better stress increases.	B has a normal session.
15:30-16:00	Wall Street opens down and A sells his contracts; A cannot fail again.	Wall Street opens down. B operates without the real-time estimation of stress in the market perception.
16:00-17:35	Wall Street starts to up strong. A had failed again. A has to buy again at a higher price.	Wall Street starts to up strong. B is operating in a normal way.
RESULTS	A has lost a lot of money. His/her emotional state has deteriorated.	They depend on the trader and on B intuition.

Table 2 Trader C's bad day vs trader D's normal day, both with AMI support

Time	Trader C Bad Day	Trader D Normal Day
9:00 -9:05	Pre-open Market. C manages many contracts, and C is stressed. Besides, C had an argument with his/her partner. The biometric sensors indicate C, that his/her emotional state is a little different of his office mates.	Pre-open Market. D manages many contracts, and D is stressed
9:05-13:00	The session is quiet, but C is conditioned by the argument and the biometric sensors alert C that C is in an abnormal state. C stops the trading process and phones his/her partner.	The session is quiet, and D has a normal trading process.
13:00-13:05	An important data is published. After the break and the conversation with his/her partner, the biometric signals indicate that C is in a normal state. Besides, C can see the stress of the market on real time.	D knows the stress state of the market in this moment, and D takes decisions with the information on real time.
13:05-17:35	C has a normal trading process.	D has a normal trading process.
RESULTS	It depends on the trader.	It depends on the trader.

gather that vital information in real time with currently available devices? The next section explains how this is possible.

5 How to Build a Trading Support AmI

Section 2 already referred to devices that can gather biometric data related to stress: Heart rate, Blood pressure, Breathing rate, Brain waves, Muscle tension, Skin resistance and Temperature. Table 3 shows some sensors that can be used to make an experiment setup to measure the stress of the trader.

Depending on the trader some sensors could be more suitable than others for market transactions. Heart rate, Blood pressure, skin resistance, muscle tension can be non-invasively sensed with existing technology. Ambient Intelligence is in the position to put extra valuable information in the focus of the traders. Biometric data can be shown to the trader in the same way they are shown other financial data in real time. With a simple trigger in the screen, the trader could be informed that his state is altered and decision-making degraded.

Table 3 Suitable Sensors for measuring the trader's stress

Sensor	Figure	Developer	Size	Details
Polar F4 Heart Rate Monitor		Polar	Fit to Chest	60 – 100 pulses as normal range
Tiny Blood pressure sensor		Fraunhofer- "Hyper-IMS"	1mm	70-110mmhg as normal range
Breath Rate The Capnocheck		BPI	Fit to finger	9-14 per minute as normal range
Brain Waves Electrode Cap		Biopac Systems	Fit to head	Stress could throw the frequency to the higher beta range
Muscle tension		Fraunhofer Institut	Embedded in an item of clothing	The greater the stress, the more likely the muscles are to produce a synchronous twitching effect.
Skin resistance SC Wrist		Philips	Wrist	When the resistance decreases the person is sweating slightly more and is more stressed
Temperature	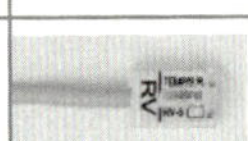	Bio-Medical	1 cm	If the hand is cooler than the neck, you could be stressed.

A clear example of this measures applied to the trader is the Rationalizer (www.design.philips.com/about/design/designnews/pressreleases/rationalizer.page) created by Philips and the bank Abn Amro (Fig.3). It consists of two components, the EmoBracelet and the EmoBowl. The bracelet measures the arousal component of the user's emotion through a galvanic skin response sensor. This arousal level is rendered as a dynamic light pattern either on the EmoBracelet itself or on the

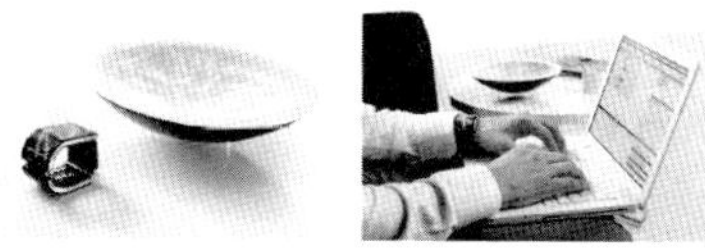

Fig. 3 Rationalizer

EmoBowl. The higher the arousal level, the more intense the dynamic light pattern becomes: the number of elements increases, the speed increases and the colour shifts from a soft yellow, via orange, to a deep red. However, the device cannot be bought and there is not available information of what measures it takes.

6 Conclusions and Future Work

Traders have complex decision making support: e.g., technical analysis, fundamental analysis, and expert systems indicating propitious moments of operation. Still, there is substantial information missing which has a very important effect in daily decisions to buy or sell stock. Market psychology is very influential in the evolution of the market, still traders do not have that data available in real-time. A trader's greatest enemy is himself and will suffer heavy losses if carried away by stress. In those circumstances avoiding making decisions could be the best decision. Equally, being aware of the stress of other traders in real time completes the vision they have about the world they operate in. Ambient Intelligence opens up a new way to support traders by detecting when a trader transitions into a non secure state for decision-making. We have shown there is a good technological offer which makes this development feasible. We have currently chosen an advanced smart-wearable device to measure levels of stress and we are in the process of testing it along with the decision process that will inform the traders.

Acknowledgements. This work has been partly funded by the Spanish Ministry of Industry under the project OSAMI (TSI-020400-2008-114), and the Spanish Ministry of Education under the projects InCare (TSI2006-13390-C02-01) and Artemisa (TIN2009-14378-C02-02).

References

1. Kowalski, K.M., Scharf, T., Vaught, C.: Judgment and decision making under stress: an overview for emergency manager. International Journal of Emergency Management 1(3), 278–289 (2003)
2. Adya, M., Phillips-Wren, G.M.: Risky Decisions and Decision Support - Does Stress Make a Difference? In: Proceedings of JAIS Theory Development Workshop, Article 55, vol. 9 (2009)
3. Delgado, M.R., Porcelli, A.J.: Acute Stress Modulates Risk Taking in Financial Decision Making. Psychol. Sci. 20(3), 278–283 (2009)

4. Augusto, J.: Ambient intelligence: The confluence of pervasive computing and artificial intelligence. In: Intelligent Computing Everywhere, pp. 213–234. Springer, Heidelberg (2007)
5. Granaham, T.: High Finance: An IDD Web Exclusive, Investment Dealers' Digest (2007) `http://dealbook.blogs.nytimes.com/2007/12/21/drugs-and-todays-wall-street/`
6. Coates, J.M., Herbert, J.: Endogenous steroids and financial risk taking on a London trading floor. Proceedings of the National Academy of Sciences, National Acad. Sciences 105(16), 6167–6172 (2008)
7. Frazier, T.W., Steinhauer, S.R., Strauss, M.E.: "Respiratory sinus arrhythmia as an index of emotional response in young adults. Psychophysiology 41, 75–83 (2004)
8. Sakuragi, S., Sugiyama, Y., Takeuchi, K.: Effects of laughing and weeping on mood and heart rate variability. Journal of Physiological Anthropology and Applied Human Science 21, 159–165 (2002)
9. Blangsted, A.K., Fallentin, N., Hjortskov, N., Lundberg, U., Risseń, D., Søgaard, K.: The effect of mental stress on heart rate variability and blood pressure during computer work. Eur. J Appl. Physiol. 92, 84–89 (2004)
10. Barnea, O., Shusterman, V.: Analysis of Skin-Temperature Variability compared to Variability of Blood Pressure and Heart Rate. IEEE Engineering in Medicine and Biology Society 2, 766–771 (1995)
11. Lang, P., Greenwald, M., Bradley, M., Hamm, O.: Looking at pictures: Af-fective, facial, visceral, and behavioral reactions. Psychophysiology 30, 261–273 (1993)
12. Adams, S., Alkon, A., Boyce, W.T., Chesney, M., Chesterman, B., Cohen, F., Folkman, S., Kaiser, P., Tschann, J.M., Wara, D.: Psychobiologic reactivity to stress and childhood respiratory illnesses: results of two prospective studies. Psychosom. Med. 57, 411–422 (1995)
13. Ossebaard, H.C.: Stress reduction by technology? An experimental study into the effects of brainmachines on burnout and state anxiety. Appl. Psychophysiology Biofeedback 25, 93–101 (2000)
14. Lori, G.: A micro simulation of traders activities in the stock market: The role of heterogeneity, agents interaction and trade frictions. Journal of Economic Behav-iour and Organization 49, 269–285 (2002)
15. Baker, M., Wurgler, J.: Investor Sentiment in the Stock Market. Journal of Economic Perspectives 21(2), 129–151 (2007)

Awareness Models for the Development of Ubiquitous Systems

Jose Figueroa Martinez, F.L. Gutierrez Vela, and Cesar A. Collazos

Abstract. Ubiquitous systems are highly dependent on context information in order to provide effective interaction for their users. The Awareness is a key property for interaction but difficult to support. The uses of conceptual models for define Awareness requirements of users and the system itself can raise the quality of the system in almost all it aspects. In this work we propose a set of models that let developers define the required Awareness for the systems and users at requirement level, in order to reduce the problems related with its support, allow measurement of quality of the system through requirements and allow reusing of models and knowledge.

1 Introduction

The ubiquitous systems are responsible to give functionality to devices that look to support the features described by Mark Weiser in his work about "machines that fit the human environment" [9]. Mobility, intelligence for interact with users and environment, invisibility, adaptation, pro activity, among others.

Jose Figueroa Martinez
GEDES, University of Granada, Avda. Periodista Daniel Saucedo Aranda s/n,
18071 Granada, Spain
email: jfigueroa@ugr.es

F.L. Gutierrez Vela
GEDES, University of Granada, Avda. Periodista Daniel Saucedo Aranda s/n,
18071 Granada, Spain
email: fgutierr@ugr.es

Cesar A. Collazos
University of Cauca, Department of Systems,
Faculty of Electric and Telecommunication Enginery, Popayan, Colombia
e-mail: ccollazo@unicauca.edu.co

J.C. Augusto et al. (Eds.): ISAmI 2010, AISC 72, pp. 237–245.
springerlink.com

The ubiquitous computer devices are made to support the paradigm "1 person: N computers", working together to help users meet their day to day goals, in the professional or personal scope.

The ubiquitous systems require being sensitive to context in order to interact with it and adapt to its constant changes. Any context adaptation requires an updated knowledge of the context. This is known as Context-Aware computing [1]. Also, ubiquitous systems act like intermediaries between user and environment, giving update knowledge to users during their decision making process.

The Ubiquitous systems cover mobile systems and share common context elements like Location [1]. However, it can work with contexts that are more complex to describe, like the emotional situation of the nearest persons to it.

These features make the development of ubiquitous systems a highly complex task [6] that begins with an extensive analysis of the information requirements from the user and the system. Later, these requirements must be supported by the system in a correct and justified way, because fulfill the general objectives depend on that.

In this work, we present a set of conceptual models that we have developed for defining the Awareness support in a computer system, at requirements level. This contribution fills the gap between conceptual models, frameworks and tools for supporting Awareness, providing a general and reusable way for defining what type of Awareness support and how/where/when to do it.

2 Awareness

The term "Awareness" can be ambiguous and confusing [8]. It is seen as one of those English words that can mean many things. In a general way, Awareness means "*the knowledge of what is going on*" [5]. For our work we use the definition proposed by Mica Endsley, which define Situation Awareness, but at the same time, describes the Awareness in the more general way:

"*Awareness is the perception of the elements in the environment, within a volume of time and space, the comprehension of their meaning, and the projection of their status in the near future*" [5].

There are other definitions of the Awareness term, but generally linked to some adjective or particular use, like Informal Awareness, Concept Awareness, Task Awareness, Context Awareness, etc. [3,4].

Perception, Comprehension and Projection are the 3 components of Awareness as an updated knowledge of the environment, including any entity [5].

The "elements of the environment" refers to any conceptual element that can be perceived by the entity that generate Awareness of that element or elements. For example, a person can acquire Awareness of the Location and the Identity from others, or their voice volume, their way to speak, their external temperature, etc.

We call "Awareness Type" to a set of elements from the environment. That is, a particular type of Awareness is formed by the perception of a specific set of elements from the environment, the comprehension and projection of their status.

In order to support Awareness, the development of a computer system must integrate those concepts in the process at different levels, beginning with the requirement analysis. This task has been addressed by other researchers like Carl Gutwin, which developed a theoretical framework, a group of tools and techniques for support Workspace Awareness in collaborative systems [6]. In the same manner, Kirsh-Pinheiro developed a framework, at design level, for support past event Awareness in Groupware systems [7].

The Awareness can be supported and implemented using these tools among others. However, there remains the need of defining what Type of Awareness are going to be supported and how are they going to fit in the current domain. It does not matter how many types of Awareness the system are going to support, the system and the developers need to know the details of its support.

In order to improve the Awareness support in this area, we present the models we developed for defining Awareness requirements and its usage details.

3 Awareness Models

The models we propose allows to include the Awareness in the beginning of the requirement analysis phase, where Awareness requirement presents the knowledge needed by domain entities for decision making in order to reach the final system goals.

The first model that we present is generally used for define Awareness Types and it is not linked to any environment. The other three models allow to define the Awareness data required on a specific domain and where are going to be used.

3.1 Awareness Type Model

This model allows defining generic Awareness Types. That is to say, the conceptual Awareness Types that can be used in any computer system development.
The structure of this model is show in Figure 1.a. This figure shows two entities: AwType and AwElement.

The AwType element represents the Awareness Type, which is composed by many conceptual elements (AwElement) from the environment or other entities. Also, it can be specialized from other Awareness Types, inheriting their structural elements, or can be composed by other Awareness Types, in order to encompass them.

The Figure 1.b shows an example of using the Awareness Type Model. This model defines two Awareness Types: Identity Awareness, composed by the element Identifier, and Location Awareness, which is a specialization of the Identity Awareness with one more element, the Location.

There are larger and more complex Awareness Types, like the Workspace Awareness [5] or Activity Awareness [2]. This Awareness Types can be seen like compositions of other simpler Awareness Types. The composition of many Awareness Types allows generating an integrated and global knowledge of the environment and any perceived entity.

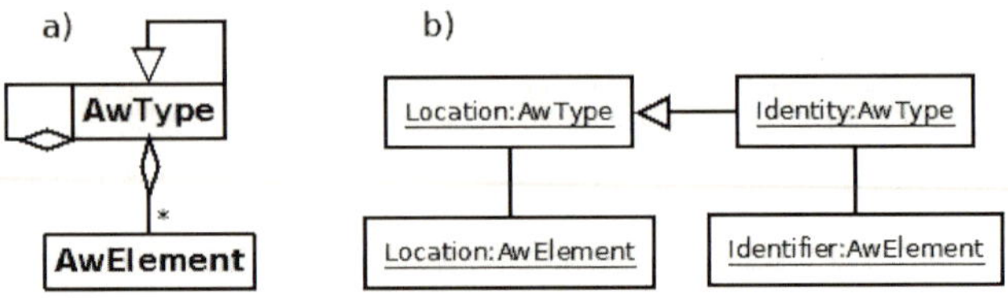

Fig. 1 a) Awareness Type model. b) Location and Identity Awareness modeling example.

3.2 *Specialized Awareness Type Model*

This model allows defining an Awareness Type in a specific computer system with a specific domain and entities. That is to say, allows representing the structure of a general Awareness Type (AwType) with a structure of entities belonging to the working domain, as shown in Figure 2.b. This is done by linking the elements (AwElement) of the used Awareness Type with the properties (SourceProperty) of the entity data source (AwSource).

The SpecializedAw represents the integration of a general Awareness Type with the domain, creating a data source from the perceived AwSource data. This is done by defining the relations between the AwElement and the SourceProperties.

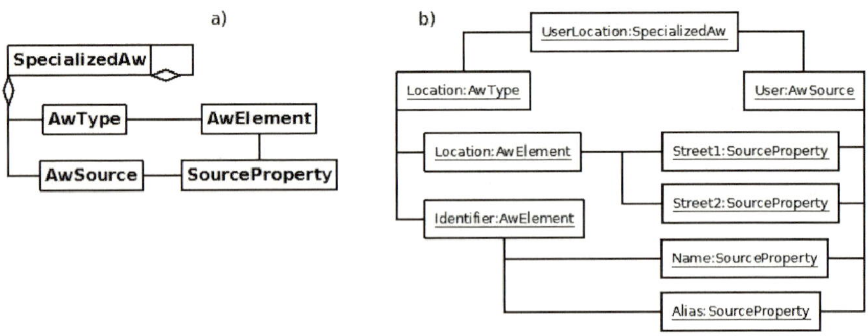

Fig. 2 a) Specialized Awareness Type model. b) Location Awareness linked to a domain entity.

An example of the SpecializedAw is shown in Figure 2.b. That model links the Location Awareness with an entity from the working domain, the User. This forms the UserLocation entity, which will work like an Awareness data source.

This model does not show how to select instances from the AwSource. That is the purpose of the Awareness Requirement Model.

3.3 *Awareness Requirement Model*

This model allows relating three things: domain entities, Specialized Awareness, Awareness Level and the constraints for select each one. This relation describes the Awareness at requirement level, introducing the necessary components that let developers to define Awareness requirements.

The structure of the AwRequirement, presented in Figure 3.a includes the AwUser which represents the domain entity that need the Awareness data, represented by the SpecializedAw. The Awareness data are going to be presented to the AwUser at some level, which can be perception, comprehension or projection and maybe some sub levels, depending on what is needed by the users and supported by the system. The different constraints allow selecting the instances of each entity in order to define a set of consumers, providers and levels of data presentation.

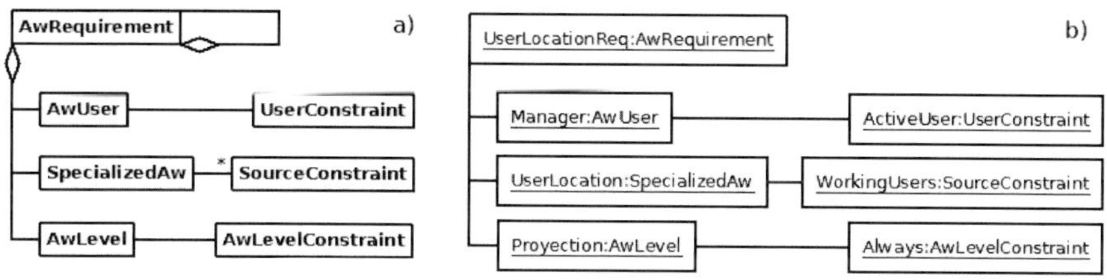

Fig. 3 a) Awareness Requirement model. b) Awareness Requirement modeling example.

The example presented in Figure 3.b shows the UserLocationReq, which represents a requirement of Awareness data needed by the active Manager. The Manager requires knowing the location of the active users at level of projection, which means that he need to know where the users are going to be in near time. The projection level must be supported "always" in this requirement.

The Awareness data can be transmitted to the AwUser through the user interface or using signals, depending if the AwUser is a human or other system.

To support the higher levels of Awareness, the system must process previously the Awareness data in order to generate new information that could help comprehension and/or projection of the perceived elements.

The usages of the Awareness data expressed on these requirements are not defined yet. For that purpose, we present the Awareness Usage Model.

3.4 Awareness Usage Model

This model allows linking the Awareness requirements with the models used during the development or with domain entities where the requirements are generated. Some possible entities can be tasks, processes, use cases, states, workflows, elements of the system modeling under different views.

The Figure 4.a shows the structure of the AwUsage, which presents a set of relations between a Domain Element, an Awareness Requirement with their constraint of usage, and a Documentation element. This model can serve to locate the awareness requirements in the development and to justify its existence/necessity.

The Figure 4.b shows the UserLocation Awareness usage, linking the MonitoringTask with the UserLocation Awareness always present, documented by the UserLocationDoc instance.

These models allow defining the Awareness data sources, "how" to integrate it in the computer system, "who" need it, at "what" level and "where" is needed. This covers the basic support for Awareness in a computer system.

This way of defining the support for Awareness in a computer system allows to manage two important and complex aspects present in every system that support Awareness: privacy and disruption. The Awareness data is provided only to entities that it requires and only at level that it is needed [4].

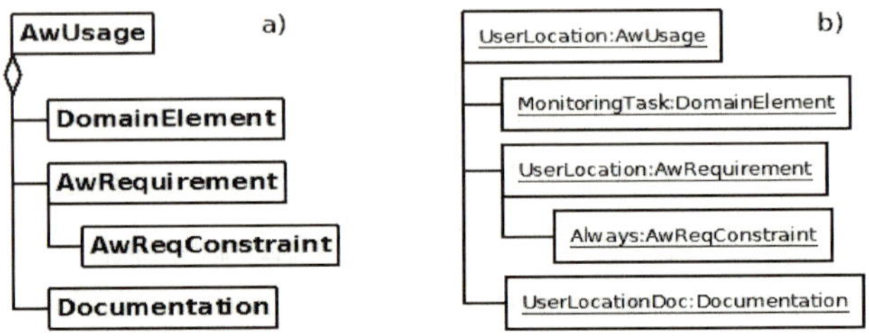

Fig. 4 a) Awareness Usage model. b) Awareness Usage example.

4 Modeling Examples

In order to exemplify the Awareness requirement and usage models we are going to describe a fictitious scene, where is required to develop an integrated system to support emergency management teams. This system is composed of many software modules, distributed on servers and ubiquitous devices carried by each member of the teams, as well as a device carried by the team leader of each team. In this case we focus some updated data requirements needed by the team leader, the active members and the artificial agents that makes work the devices.

Table 1 Awareness requirements definition

AwReq	AwUser[a]	SpecializedAw[b]	AwLevel[c]
TeamLoc	TeamLeader of every working group.	TMLocationAw of every members of his team.	Always simple perception.
HostHealth	Ubiquitous Health Agent of every active worker.	HeatlhAw of host members.	Always basic comprehension and perception.
NeiborsLoc	Ubiquitous Location Agent of every active worker.	LocationAw of the nearest Location Agents.	Always perception.

[a]and UserConstraint. [b]and SourceConstraint. [c]and AwLevelConstraint.

These devices provide *off line* functions for locate other members of the team. That is, they work even without communication with the operation base. Also, one of them monitors the health condition of the person. In Table 1 are defined some Awareness requirements of the described entities.

The Awareness Requirement TeamLoc can be textually expressed as "The TeamLeader of every working group need the TMLocationAw of every member of his team at perception level". As shown, the requirement answers the questions of *who*, *what* and *how*. The question *when/where* is answered by through the Awareness Usage Model.

The Awareness Requirement HostHealth can be textually expressed as "The Ubiquitous Health Agent" of every active worker need the HealthAw of his carrier user at basic perception and basic comprehension." Here, the Awareness level expresses that is needed a basic perception of health, like heart rate. Also, it is needed to know, with the help of some data preprocessing, things like "possibility of heartbreak". That depends on the capabilities of the system and health sensors. The word "always" express that the level of Awareness that is needed in this requirement, which is perception and basic comprehension, not going to change.

The modeling element or DomainElement shown in Table 2 consists on systems tasks. In this case, we present the IncidentResolving and TeamMovement tasks.

IncidentResolving represent the state of the systems during the task of resolving an incident. For example, in case of fire, the task begins with the arrival of the teams to the fire, and ends when the teams left the area.

TeamMovement represents the task of team mobilization to the incident area and vice versa.

Table 2 Awareness requirements usage

DomainElement	AwReq[a]	Documentation
IncidentResolvingTask.	TeamLoc.	During the incident, the TeamLeader has to distribute tasks between the members.
IncidentResolvingTask.	MemberLoc.	During the incident, it is needed by every member to know where are the other members in order to work together and help each other.
IncidentResolvingTask, MovementTask.	HostHealth.	To monitor the health of the members is needed to safeguard life of the members.

[a]and AwReqConstraint.

The Table 2 is based on the Awareness Usage model, describing the modeling elements on which the Awareness requirements are used, the constraints of usage and the appropriate documentation to justify it. Moreover, this model allows locating the reasons that originated the implementation of the Awareness mechanisms supported by the system, limited only by the quality of the documentation.

The models representation is not limited by some technology. In this example we used a table notation, because the model's structure allows that flexibility.

During development phases, these requirements are going to generate modeling elements that are going to traduce in Awareness mechanisms incorporated on implementation elements of the system (user interfaces, data gathering, etc.).

5 Conclusions and Future Work

The presented models allow defining the Awareness support in a computer system at requirements level. This support is necessary in ubiquitous systems. Although these models work at conceptual level, their usage lets to validate and justify or even generate, at basic level, the Awareness support on the system. Moreover, their allow reusing the requirements knowledge in other developments.

Other advantage of these models is that can be used for modeling the requirements of context data needed by the system for adaptation to the human behavior. Another advantage is the integration of the Awareness support from the requirement analysis of the system, letting developers to work in what is needed and justified. This raises the quality of the system and helps their maintenance.

Privacy, disruption and information overload are managed implicitly. However, this problem must be analyzed during the development phased, which can see the benefits of include Awareness information on the models. In addition, the system can implement adaptation mechanisms in order to reduce disruption. Again, these models can used to define the Awareness required for that adaptation.

As part of the future work, we find feasible the creation of a catalogue of Awareness Types for gathering general knowledge about distinct Awareness and usages.

We think that other important use of the models is to generate parts of the system from the defined requirements. Can be generated user interfaces, scaffoldings, validation tests, quality tests, etc. For this task are required other models at design level, which let to define more specific features in the system.

The Awareness is key for context aware applications. Its correct support is necessary for the successful of the ubiquitous systems. Its effective management during development is essential for reduce their management complexity.

Acknowledgments. This research is financed by the Ministry of Science and Innovation, Spain, as part of DESACO Project (TIN2008-06596-C02-2) and the Mexican National Council of Science and Technology CONACYT.

References

1. Abowd, G.D., Mynatt, E.D.: Charting Past, Present, and Future, Research in Ubiquitous Computing. ACM Trans. Comput.-Hum. Interact 7(1), 29–58 (2000)
2. Carroll, J.M., et al.: Awareness and teamwork in computer-supported collaborations. Interacting with Computers 18(1), 21–46 (2006)
3. Dourish, P., Bellotti, V.: Awareness and Coordination in Shared Workspaces. In: Proc. of the 1992 ACM conference on Computer-supported cooperative work, pp. 107–114 (1992)
4. Drury, H., Williams, M.G.: A Framework for Role-Based Specification and Evaluation of Awareness Support in Synchronous Collaborative Applications. In: Proc. of the WETICE, pp. 12–17 (2002)
5. Endsley, M.R.: Towards a Theory of Situation Awareness in Dynamic Systems. Human Factors 37(1), 32–64 (1995)

6. Gutwin, C., Greenberg, S.: A Descriptive Framework of Workspace Awareness for Real-Time Groupware. Computer supported cooperative work 11(3), 411–446 (2002)
7. Kirsch-Pinheiro, M., et al.: A Framework for Awareness Support in Groupware Systems. In: Proc. of the 7th Int. Conference on CSCW in Design, pp. 13–18 (2002)
8. Schmith, K.: The Problem with Awareness. Introductory Remarks on Awareness in CSCW. Computer Supported Cooperative Work 11(3), 285–298 (2002)
9. Weiser, M.: The computer for the twenty-first century. Sci. Am. 265, 94–110 (1991)

A Statistical-Relational Activity Recognition Framework for Ambient Assisted Living Systems

Rim Helaoui, Mathias Niepert, and Heiner Stuckenschmidt

Abstract. Smart environments with ubiquitous sensing technologies are a promising perspective for reliable and continuous healthcare systems with reduced costs. A primary challenge for such assisted living systems is the automated recognition of everyday activities carried out by humans in their own home. In this work, we investigate the use of Markov Logic Networks as a framework for activity recognition within intelligent home-like environments equipped with pervasive light-weight sensor technologies. In particular, we explore the ability of MLNs to capture temporal relations and background knowledge for improving the recognition performance.

1 Introduction

Simple daily tasks are usually accomplished effortlessly. However, solving them might present a hard struggle for elderly people and individuals with disabilities or cognitive impairments. Due to the rapid population aging, there is an urgent need for more efficient methods towards reliable and continuous healthcare systems with reduced costs. Smart environments with ubiquitous sensing technologies are a promising perspective in this context. A primary challenge for such Ambient Assisted Living (AAL) systems is the automatic recognition of the activities carried out by the human at their domicile. This allows to anticipate and to provide the needed services at home and in time. Current approaches to this problem can be classified according to (1) the learning paradigms used, (2) the relevance of temporal information in the recognition model, and (3) whether the model is linked to real sensor data or simulated data. The current main learning paradigms applied are roughly

---

Rim Helaoui · Mathias Niepert · Heiner Stuckenschmidt
KR & KM Research Group, Computer Science Institute, University of Mannheim, Germany
e-mail: {rim,mathias,heiner}@informatik.uni-mannheim.de

J.C. Augusto et al. (Eds.): ISAmI 2010, AISC 72, pp. 247–254.
springerlink.com

data-driven or knowledge-based. The work presented in [2] and [13] are two examples that apply knowledge-driven learning algorithms. In their models, a prior rule base is mandatory to react to the incoming information. A formal ontology of the user's intentions is used in [13] to enrich the current domain knowledge und reduce uncertainty. Some recent work emphasizing the importance of the temporal context proposes a novel approach to recognize human activity with constraint-based temporal reasoning framework (OMPS) [8]. Starting from general predefined rules, the system's procedure performs a search in the space of Decision Networks for a set of synchronizations that is applicable given the current state of the Decision network [8]. However, to obtain high level activities, the system resorts to heterogeneous sensor data like a stereo camera on the ceiling with a tracking system. The use of such tracking systems implies some privacy and costs issues which we would like to avoid.

Most of the current approaches to our problem are, nonetheless, data-driven. However, they seldom integrate temporal constraints in their systems. The majority use supervised learning methods like Hidden Markov Models (HMM) to recognize human activities. Obtaining substantial amounts of labelled data is often a bottle-neck for these approaches [3]. On another side, recent efforts to explore unsupervised learning methods have been developed. In [4], web mining is used to automatically extract the probabilities of the co-occurrence of some activities and the used objects. These probabilities assemble a HMM capable of recognizing activities in traces of object data. Increasing interest can be noticed concerning weakly supervised learning like multi-instance learning [3] and semi-supervised learning methods. We conjure that combining both paradigms, i.e. data-driven and knowledge-based, is an appealing direction to solve such non-Markovian problems and capture a rich temporal context. Similar to [5], our work also suggests to use both probabilistic models and relational information to transform the raw sensor data into higher-level descriptions of people's behaviours and activities. In [5] all of the ontology-based methods, web mining and HMM are used to produce a richly structured dataset describing people's daily patterns of activities. Due to the limited HMM temporal dependencies, however we opt for Markov Logic Networks (MLN). Indeed, MLN does not restrict the temporal context to some predefined range and can reason with variably long user's behavioral history. Furthermore, both logical statements as well as probabilistic ones are easily united in one single framework. This allows, for example, to flexibly integrate the existing domain knowledge, which reduces the amount of training data needed(see Section 2). To summarize, many existing approaches rely on manually constructed sets of rules for recognizing activities and/or rarely take temporal knowledge into account. This makes successful recognition of interrupted activities and activities engaging multiple actors rather out of reach. The purpose of this work is to investigate the use of MLN as a framework for activity recognition based on information from pervasive light-weight sensor technology. In particular, we want to test the ability of MLNs to capture temporal relations and background knowledge for improving the recognition performance. To this aim, we address daily human activities within intelligent home-like environments equipped with pervasive light-weight sensor technologies [14].

2 Markov Logic Networks for Activity Recognition

A Markov logic network combine first-order logic and undirected probabilistic models. It is essentially a set of weighted first-order logic formulae. The formulae and their weights can be automatically learnt as proposed in [16] for instance. The real-valued weights represent the confidence one has that each of these formulae hold. Probabilistic inference is performed over the joint probability distribution over all ground atoms. The formulae can be divided into soft and hard constraints. Hard constraints are required to always hold whereas soft ones can be violated in a solution. These weights can be learnt if training data is available. Thus, soft constraints will model the uncertainty of the events and the corresponding activities. Two examples of such soft rules are depicted below, where the predicates *activity* and *sensor* model daily activities and sensor events, respectively, at timestep t.

$$\forall timestept : activity(settingthetable, t) \rightarrow activity(eating, t+1) \tag{1}$$

$$sensor(Oven, t) \rightarrow activity(cooking, t) \tag{2}$$

Hard constraints allow integrating existing common-sense knowledge, which can reduce the amount of needed training data. An example of such a hard constraint would be that a person cannot eat while sleeping. It can also express relevant physical and temporal constraints like the fact that a person cannot be present at two different places at the same time. Unlike other modelling methods such as HMM, MLN do not require a predefined number of states. They also offer the flexibility to reason with variably long user's behavioral history. Thus, it permits to cover different temporal contexts and explore its effect on the system's recognition and prediction accuracy. MLNs are applied for activity recognition in [6] and [7]. Unlike our framework, however, both approaches rely on video data. Compared to [6], we aim at an expanded set of more complex daily activities, addressing concurrent ones, as well as a richer temporal context. As proposed in [7], missing sensory data is an interesting aspect to be handled. Exploiting common-sense knowledge for a better model is another similarity to our proposed work. The authors, yet, limit their work to simple activities with no explicit consideration of the temporal context.

3 Activity Recognition Framework

The long term goal of this work is a framework towards proactive ambient healthcare systems. The approach applies MLN and relies on lightweight sensor data with real-world deployment. We expect our system to offer a good opportunity to explore the effect of different temporal contexts on the system's recognition and prediction accuracy. We intend to extend the framework to the case of more than one actor and to investigate the viability of the system under limited availability of sensor data. The system can eventually be applied for the case of anticipation and prevention of certain diseases based on behavioral-symptoms such as depression [14]. To achieve these expected contributions, many challenges have to be addressed. Indeed, activities of daily living are usually non sequential. They are often overlapping, alternating or even interrupted. Moreover, most of them share a large subset of sensors and objects. The activation of these sensors do not always follow the same temporal pattern for the same activity. We propose to integrate background knowledge covering many aspects. These can be categorized in three dimensions: Information about the user, about the environment and about time. Table 1 offers a refined overview of the model's background knowledge. In the proposed framework, low

Table 1 Background knowledge for activity recognition using MLN

User	Environment		Time
	Indoors (sensors)	Outdoors	
ID	ID	Weather	Timestamp
Location	Type	Temperature	Part of the day
Mental state (drives[a])	State		Day of the week
			Season of the year
			Holidays and special occasions

[a] The drives correspond to natural needs like sleep, hunger. The sensor's states reflect whether a sensor is on or off for example.

level signals are received as periodic events with a timestamp. Those and the available background knowledge (see Table 1), are described using the corresponding model's predicates. The low level events could be aggregated into actions such as "using the oven" or "entering the bedroom" as examples. The well defined activities of daily living, which we intend to recognize, are then inferred at a higher level. To address the temporal context, we propose to cover both of the qualitative and quantitative aspects. For the latter, we can introduce time intervals of the different events, actions and activities, which also captures their duration. A tolerance range can also be assigned to the start- and end-time. An average window-length can be calculated for each activity and be used as an additional feature during the recognition process [9]. Another option would be to cluster the different durations into classes like "very short, short, long and very long" instead of giving an explicit value. Two components, which can be either an event, an action or an activity, can be distinct, overlapping or alternate. The latter case is a special one of distinct components. To capture those possible relations between two intervals, i.e. the qualitative temporal aspect, the framework can be extended with a subset of the thirteen Allen's base relations. Indeed, for distinct activities we propose to use the relations: "before", and "meets". For the overlapping activities we limit our relations to "overlaps" and "during" (see [10]). Covering the temporal dimension is not only expected to help improve the expressiveness of the model and its recognition accuracy but could also be a good information for some user's states. Indeed, the rapidity of increase in carrying out some activity could be an indicator of its urgency. As already mentioned, there are events, actions and activities that entail the need to assert decisions on other components based on predefined common sense rules as well as physical and temporal constraints (see above). These relationships must be synchronized (propagated) to be kept compatible with the recognized activities. A simple instance would be that the activity "clear the table" entails that "set the table" must have taken place anteriorly. Hard constraints are a good option to express them in our framework. Hence, MLN use those predefined hard constraints and the obtained weighted rules to infer the current activity (see 2). As a beginning to our work, we decided to use a reduced set of sensors that only label some objects in the assisted living home. The activities to be recognized are restricted to one at a time. We do not address the qualitative temporal aspect of the actions and we only consider two temporal qualitative relationships, namely *next* and *after*: For two activities, actions, or events a at a timestamp t and b at timestamp d:

$$t+1=d \Leftrightarrow next(b,a)\ ;\ t+1>d \Leftrightarrow after(b,a) \tag{3}$$

3.1 Experiments

In this section we present some preliminary experiments verifying the viability of the proposed approach for activity recognition. We defined two predicates to describe the events and the activities. These are *sensor(sensor-id, timestamp)* and *act(activity-id, timestamp)*. The possible model uses two soft formulae:

$$\forall s_1,a_1,t : sensor(s_1,t) \rightarrow act(a_1,t) \tag{4}$$

$$\forall a_1,a_2,t\ : act(a_1,t) \rightarrow act(a_2,t+1) \tag{5}$$

The training data consists of the periodic events labelled with their corresponding activities [1], which are considered as the item sets of our rules mining. The sensor-activity and the activity-activity pairs having a given support are extracted and related, as weights to the rules defined above. Formula 4 models the conditional probability of activity a_1 given sensor s_1. Formula 5, on the other hand, captures the probability of activity a_2 being carried out at time step $t+1$ given that activity a_1 occurred at time step t. Since we restrict our activity recognition to one activity at a time in those experiments two more hard constraints have to be added. In the first experiment, the first hard constraint exclude parallel activities. The second rule expresses that at least one activity should take place at each time step. Considering sensor data as observation and the current activity as hidden in our MLN, the system reasons based on the current state as well as the previous ones.

3.2 Dataset and Setting

To train the described system, we considered the publicly available dataset RFID Data[1]. This was collected in a lab outfitted with 60 RFID tag, placed on the different objects involved in performing the set of eleven fine-grained activities: "Using the bathroom", "Making oatmeal", "Making soft-boiled eggs", "Preparing orange juice", "Making vanilla latte", "Making tea", "Making or answering a phone call",

"Using the door", "Setting the table", "Eating breakfast" and "Clearing the table". The user wore two RFID gloves that detected the RFID tag within 2 inches. The collection covered 30 min per day, during ten days. The data is provided in form of timely ordered events relating the sensor ID and the activity being carried out for each timestamp. For the implementation of the proposed activity recognition framework, we used "TheBeast" [15]. The weights of the soft formulae are pre-computed to ensure a faster recognition process. This was achieved by applying an efficient algorithm for learning association rules [11]. Since we tried to extract as much itemsets as possible, we provided very low support and confidence values of 10^{-9} and 10^{-2}, respectively. The MLN inference task has been performed by the ILP solver SCIP [12]. The first day was kept as test data and the remaining ones was used to train the system.

3.3 Results

Our system has been evaluated using precision, recall and the F-measure. First, we only modeled the "next" temporal relationship and used a simple set of two soft rules(see 4 and 5) along with two hard constraints. Those ensure that at each timestamp exactly one activity takes place. An intuitive consequence is that the three mentioned measures will coincide. To avoid redundancy, we only mention the precision value while reporting our results. This first model reached a value of 0.77. The system tried to optimize over the whole test data with a combination of more than 1300 time steps, 10 activities and 59 sensors. Since we are more concerned with the recognition of current activity, we applied time-windows with some length n to our test data. Thus, the model infers using the last n events which, we claim, can alleviates and improve the recognition accuracy, even it reduces the available context. The length of the time window has been chosen experimentally after comparing different values. We fixed n to roughly twice the mean length of an activity, i.e. 160 time steps. The accuracy results with the corresponding activities are depicted in Figure 1. A significant accuracy deterioration is observed starting from the $5th$ window. This corresponds to activities sharing a large set of common sensors. Indeed, "setting the table", "eating breakfast" and "clearing the table" can hardy be distinguished with provided rules. To address this ambiguity, we insert three hard constraints as a sample common-sense knowledge. Using the "$after$" temporal relationship introduced in Section 3, we state the following: (a) "clear the table" after "eating breakfast"; (b) "eating breakfast" after "setting the table"; and (c) "clear the table" after "setting the table". The model with these hard constraints raises the accuracy from 0.57 to 0.81 over the second half of the test data (see Table 2) . The overall accuracy was also improved and attained an accuracy of 0.87.

4 Discussion and Conclusion

MLN offer a high flexibility to expand and to richly describe domain knowledge. They also combine it with learnt probabilistic rules for an adaptive behaviour. Those

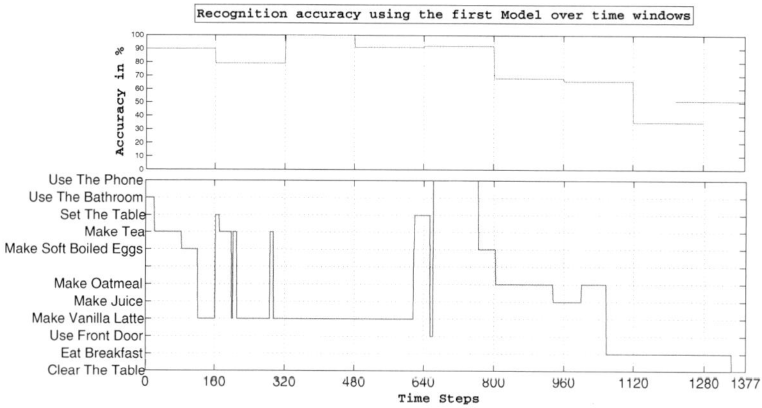

Fig. 1 Inserting time windows generally improves the recognition accuracy. A significant accuracy deterioration is observed starting from the $5th$ window. This corresponds to activities sharing a large set of common sensors: "setting the table", "eating breakfast" and "clearing the table".

Table 2 Comparing The recognition accuracy of the two proposed models. Model 1 only implements the *next* time-relationship and bases on two soft rules and two hard constraints. Model 2 expands Model 1 with the $after$ temporal-relationship encoded into three new hard constraints.

	Model 1	Model 2
Accuray over the first Time window	0.91	0.90
Accuray over the second Time window	0.57	0.81
Overall accuracy	0.77	0.87

features make them an appealing approach to activity recognition. In this work, we have presented our preliminary experiments that showed the viability of MLN for this task and their ability to capture and integrate the temporal aspect of the activity for a better inference. Our experiments also showed that the temporal context is a crucial feature that highly influences the recognition accuracy. Indeed, we proposed two simple models. The second covers more temporal relationships than the first and outperformed it over the whole data as one single time window as well as in the case of splitting the test data into two time windows. However, it deteriorated the accuracy from 0.51 to 0.4 when applied to a reduced time window covering the last 160 time steps. The activities coinciding with this time slice are those that share a large set of common sensors.

Future work will focus on a the temporal aspect including the quantitative one. We hypothesize that capturing the order of the activities into the soft rules can improve the recognition model. For example, we merely state that "eating breakfast" can not occur before "setting the table" in the current models and overlook the aspect that the activity following "setting the table" is most probably "eating breakfast". This is obviously absent in the current models since the majority of activities spread over a large number of time steps so the weights captured by second soft rule (see Rule 5) strictly cover the case that the same activity will most probably take place at the following time step. The first soft rule can also be a subject to improvement, if we cluster the sensors into classes. For example two distinct spoons bearing two different IDs can be grouped under the type spoon. This will lead to a more realistic association between activities and sensors types.

References

1. Patterson, D.J., Fox, D., Kautz, H., Philipose, M.: Fine-grained activity recognition by aggregating abstract object usage. In: ISWC 2005 (2005)
2. Augusto, J.C., Liu, J., McCullagh, P., Wang, H., Yang, J.: Management of uncertainty and spatio-temporal aspects for monitoring and diagnosis in a Smart Home. In: IJCIS (2008)
3. Stikic, M., Schiele, B.: Activity recognition from sparsely labeled data using multi-instance learning. In: Choudhury, T., Quigley, A., Strang, T., Suginuma, K. (eds.) LoCA 2009. LNCS, vol. 5561, pp. 156–173. Springer, Heidelberg (2009)
4. Wyatt, D., Philipose, M., Choudhury, T.: Unsupervised activity recognition using automatically mined common sense. In: AAAI (2005)
5. Choudhury, T., Philipose, M., Wyatt, D., Lester, J.: Towards activity databases: Using sensors and statistical models to summarize people's lives. Data Eng. Bull. 29, 49–58 (2006)
6. Biswas, R., Thrun, S., Fujimura, K.: Recognizing activities with multiple cues. In: HUMO 2007, pp. 255–270 (2007)
7. Tran, S.D., Davis, L.S.: Event modeling and recognition using markov logic networks. In: Forsyth, D., Torr, P., Zisserman, A. (eds.) ECCV 2008, Part II. LNCS, vol. 5303, pp. 610–623. Springer, Heidelberg (2008)
8. Cirillo, M., Lazellotto, F., Pecora, F., Saffiotti, A.: Monitoring domestic activities with temporal constraints and components. In: Proc. IE 2009, pp. 117–124 (2009)
9. Tapia, E.M., Intille, S.S., Larson, K.: Activity Recognition in the Home Using Simple and Ubiquitous Sensors. In: Ferscha, A., Mattern, F. (eds.) PERVASIVE 2004. LNCS, vol. 3001, pp. 158–175. Springer, Heidelberg (2004)
10. Allen, J.F.: Maintaining knowledge about temporal intervals. Commun. 26, 832–843 (1983)
11. Bodon, F.: A Survey on Frequent Itemset Mining, Technical Report, Budapest University of Technology and Economic (2006)
12. Achterberg, T.: Constraint Integer Programming, Ph.D. dissertation, Technische Universität Berlin (2007)
13. Hochul, J., Taehwan, K., Joongmin, C.: Ontology-Based User Intention Recognition for Proactive Planning of Intelligent Robot Behavior. In: MUE, pp. 158–175 (2008)
14. Helaoui, R.: Towards a Proactive System Based on Activity Recognition. In: PerCom (to appear 2010)
15. Riedel, S.: Improving the accuracy and efficiency of map inference for markov logic. In: Proc. of UAI, pp. 468–475 (2008)
16. Biba, M., Ferilli, S., Esposito, F.: Structure Learning of Markov Logic Networks through Iterated Local Search. In: ECAI, pp. 361–365 (2008)

Author Index

Printed by Printforce, the Netherlands